Natalie Isaacs is the founder and CEO of 1 Million Women, a global movement of women and girls who take practical action to fight climate change by changing the way they live.

A former cosmetics manufacturer, Natalie realised that individual action is a powerful path to solving the climate crisis. She decided to leave behind the overpackaged world of skin and beauty care to create an organisation that inspires and empowers women to act.

Under Natalie's leadership, 1 Million Women has become one of Australia's largest networks acting on climate change and is rapidly expanding internationally, with more than 800,000 women and counting. A sought-after presenter and The *Australian Geographic* society's 2017 Conservationist of Year, Natalie delivers a simple message that resonates with women and girls of all ages, the story of her journey from apathy to climate action cutting through complexity. She is a pioneer in the gender and climate change arena in Australia, and is recognised and supported by some of the world's most influential women climate leaders.

To
Dear Kate
Thank you so
much for a fantastic
time with Kate
love
Mat xx

EVERY WOMAN'S GUIDE TO SAVING THE PLANET

NATALIE ISAACS

1m♀

ABC
Books

ABC Books

The ABC 'Wave' device is a trademark of the
Australian Broadcasting Corporation and is used
under licence by HarperCollins*Publishers* Australia.

First published in Australia in 2018
by HarperCollins*Publishers* Australia Pty Limited
ABN 36 009 913 517
harpercollins.com.au

HarperCollins*Publishers*
Level 13, 201 Elizabeth Street, Sydney, NSW 2000, Australia
Unit D1, 63 Apollo Drive, Rosedale, Auckland 0632, New Zealand
A 53, Sector 57, Noida, UP, India
1 London Bridge Street, London, SE1 9GF, United Kingdom
Bay Adelaide Centre, East Tower, 22 Adelaide Street West, 41st Floor,
    Toronto, Ontario, M5H 4E3, Canada
195 Broadway, New York, NY 10007

A catalogue record for this book is available
from the National Library of Australia

ISBN: 978 0 7333 3967 7 (paperback)
ISBN: 978 1 4607 1078 4 (ebook : epub)

Cover design by Hazel Lam, HarperCollins Design Studio
Cover images by shutterstock.com
Internal Design by Renee Bahr @ Post Pre-press
Author photograph by Jenny Khan
Printed and bound in Australia by McPherson's Printing Group
The papers used by HarperCollins in the manufacture of this book are a natural,
recyclable product made from wood grown in sustainable plantation forests. The
fibre source and manufacturing processes meet recognised international environmental
standards, and carry certification.

To my children, Bronte, Jacob, Shea and Isaac,
who nourish my soul every moment of every day.

And to my beautiful granddaughter, Harper.
I'm in this fight for you, baby.

Contents

PART 2
Saving the planet: Toolkits

About 1 Million Women

A decade ago, I was inspired to create, from scratch, a new women's organisation dedicated to fighting climate change by changing the way we live.

The story of 1 Million Women is the story of individuals – **like you and me** – taking on one of the hardest challenges our modern world knows: changing our own behaviours and the way we live our daily lives. This includes the stuff we buy, the food we consume, the energy that powers our homes and vehicles, and the waste and pollution we're responsible for every day.

Here's how it all began.

Back in 2006, when I was running my cosmetics manufacturing business, I realised there was a huge disconnect in my life. There was a bunch of prompts, which I share in the chapters ahead, but in essence I was becoming conscious of climate change and the huge threat it poses to people and planet. I had this growing awareness, but I wasn't doing anything about it. I was talking the talk, but I wasn't walking the walk.

So, I decided to do something, and reducing my family's electricity use seemed a good place to start.

Within a few months, without too much effort, I'd cut our energy bill by 20 per cent. Result! I was hooked! That sparked the beginning of my climate-action journey. A journey that changed my life forever.

I could easily have stalled there. It was far from clear what I should do next. Yes, I'd taken this initial step, but where on earth do I go now? To find out, I hit the internet to see who and what was out there. I discovered a world full of organisations dedicated to the environment. While many of them were wonderful in their own way, none of them spoke to me personally. None gave me an emotional connection with what I'd just gone through.

I'd only just moved on from inaction and unawareness, and I would have loved to have found an existing organisation that said: 'Don't worry, we get it. We're just like you. It doesn't matter that you've been floating along doing nothing. Come with us now and let us show you what you can do next.'

When you take those first tentative steps to changing your lifestyle, you want like-minded people around you – not to mention a healthy dose of reassurance.

I felt there must have been millions of women like me, who sat outside the 'green belt' of activism and weren't acting because they simply didn't know what to do, or where to start. I thought that if I could tell them all that had happened to me – how easy and effective making those small changes could be, how empowered they'd feel from it and where that could lead them – they'd want to come along for the ride.

The genesis of 1 Million Women was taking shape.

I still didn't know a lot about climate change. I didn't even know the name of my country's environment minister at the time. I wasn't an activist. I'd never even marched for a cause. And I certainly didn't know how to grow a movement.

Self-doubt assailed me. Did I even have the right to be acting in this space? Who was I to tell others how to live their lives differently? But the power I felt from my own experience kept me going, as did the incredible support I received from women (and men) around me from the very beginning.

Imagine if there was a way to harness that extraordinary power each of us has so that, together, we can shape the kind of world we want to live in.

With the help of some amazing women, I began to flesh out the idea of a new movement. **1 Million Women.**

> ✅ I was just one woman who had taken action and made a difference. But I couldn't help thinking: **imagine how powerful millions of us would be.**

As with most start-ups, raising money was a huge hurdle in the beginning, but when we reached out for help, the response was there in a heartbeat. We found vital financial support as well as pro bono offerings from so many sectors – legal, accounting, PR and advertising – all of it filled with generosity and passion. Then came the ambassadors. **Sixty wonderful women, from all walks of life, became ambassadors for our cause.** They gave their time for nothing because they believed that the time was right for women to take centre stage in this climate fight.

After two-and-a-half years of planning, 1 Million Women was launched in mid-2009 at the University of Sydney. I was so nervous I could barely breathe. *What if no one turns up? What if the media don't get behind it? What if no one ever hears about us?* Walking into the venue and finding a crowd of 300 people gathered was gobsmacking. I think it was the first time I'd

seen all the people I'd been working with in the one space –
1 Million Women was *real*! And I needn't have worried. We got
lots of media coverage.

Afterwards, while a few of us were debriefing at a pub around
the corner, one of my friends checked our just-gone-live website.
'Oh my God, we've got 400 people already!' she shrieked.

Our movement had begun.

We began with a single, clear proposition. Join up and commit to
cutting 1 tonne of carbon dioxide pollution from your life within
a year, and we'll show you how. What we were saying was: when
it comes to acting on climate change in your daily life, we know
the path. We've walked the road to self-empowerment, and it's a
pretty great place to travel. **Let us show you the way.**

> ✅ The 1 Million Women message
> is specific: **we need a lifestyle
> revolution to fight the climate
> crisis, and women are the
> powerful agents of change.**

In hindsight, building our 1 Million Women community has been not so different from the way I cut that very first electricity bill: by taking one step, doing one thing, seeing an incredible and immediate result, and letting the guaranteed empowerment we felt drive us to our next step. **We can all profoundly change the way we live in exactly the same way – one step at a time.** That's how we'll make a difference. To this day, action – and easy, small steps – remains at our core.

> ✔ If there's one practical step I could tell you to do straight away, it's this:
>
> **Next time you want to buy something, stop. Take a breath. And ask yourself: *Do I really need it?***

We empower women and girls to take practical action on climate change through the way we live. Through the choices we make every single day – saving household energy and implementing clean energy options, addressing this crazy world of overconsumption, dealing with food waste and starting better food practices, getting from A to B with the least impact on the planet, supporting sustainable fashion and the economic power of women. And we show the power of our collective impact when we act together.

From the very beginning, our aim has been to appeal to people's hearts as well as heads, to move people in a way that helps them *genuinely* change the way they live. We're learning every day, and every day we're asking: *Did that work? Did we help to change behaviour? Was that event effective, or would a campaign have worked better?*

There's no doubt that the collective strength of women taking action themselves, and supporting each other along the way, makes a massive difference.

You're the voice

Before the 2015 United Nations Climate Change Conference, which lead to the historic Paris Agreement in 2016, we wanted to do something hopeful and big. Something that rallied people who weren't engaged in the whole UN process and bring it to them in an uplifting way. Something that spoke to our belief that only climate action gives us climate hope.

We landed on covering John Farnham's legendary song, 'You're the Voice', with, of course, some big voices: Wendy Matthews, Deni Hines, Melinda Schneider, Ursula Yovich and a whole bunch of wonderful women.

Everyone gave their time to pull together this beautiful project, using the power of music to convey a simple message.

It's a glorious thing to witness women, who previously haven't been engaged, move from inaction to the camp of action, and discover how empowering it is. Every day, I have the privilege of seeing women share their experiences and their advice, their achievements and their thoughts. And finding their voice, because that's what comes with empowerment: you gain confidence in what you're doing, and you find your voice.

If you're taking action yourself, inevitably you'll also feel more comfortable in demanding others take action, too, whether it's your vote at election time, the products you buy or reject, the brands you embrace or boycott, the investments you make or avoid – you become more powerful.

That's what happened to me when I took action, and it's happening to all of us at 1 Million Women.

At 1 Million Women, we talk about uplifting, empowering, inspiring, motivating and mobilising. It's not that we set out to use these words, it's just the way we've thought from the beginning – we lifted each other up and encouraged one another, reassuring ourselves, 'We can do this.' I think women are so good at this.

The way we talk to one another has always been about what we can do not what we can't, about solutions not problems, opportunities not oversights, and collective impact not individual isolation. **We've always emphasised the fact that each of us can do something and, together, we are powerful.**

There is simply no time to talk about guilt, no time to waste scolding ourselves with 'I've been doing this all wrong' and 'It's my fault.' We don't bother with those downers, despair and denial, but stay focused in that place of action.

Who wouldn't want to part of that? I'm guessing you do, or you wouldn't be here.

So, read on, and let's find out how we can all take action on climate change.

Together, we can literally save the world.

In memory of Tara Hunt, one woman who made a difference

It was December 2012, and the end of a particularly challenging year, when an email popped into my inbox.

The email was short:
Natalie, could we meet up for coffee sometime and talk about 1 Million Women? Tara Hunt.

There was something about that email; I called Tara within seconds. We had our coffee at a beachside café the following week, and that's how my extraordinary adventure with Tara Hunt began.

Tara was a special kind of philanthropist, a hands-on type who gave money but also gave of herself. 1 Million Women became the lucky beneficiary of that. Tara provided much-needed funding at a point when we needed it most. She gave us time to breathe and, together, we shaped our vision to become a global movement of millions. We discussed endlessly the state of the planet and our love of this Earth. We talked deeply about our children (Tilly, Lizey and Banjo for Tara; and Bronte, Jacob, Shea and Isaac for me) and the world we were leaving them. We strategised, planned and plotted for 1 Million Women all the time (and I mean *all the time*). We had each other's backs, and we would giggle and scheme like schoolgirls in the back row, hiding from the teacher. We loved to belly laugh and we loved each other. From the moment we met, we grew 1 Million Women together.

We shared many memories that I cherish. Like in 2013, when we took our girls to Warsaw, Poland, for that year's United Nations Climate Conference, at which 1 Million Women was receiving our first international award.

In September 2016, however, I suspected something was going on with Tara. Then she gave me the news: lung cancer, stage four. She probably had eight months to live.

She wasn't a smoker.

Tara fought cancer with courage and dignity and beauty, but she lost her battle. She passed away on 28 May 2017, aged only 54. Tara loved 1 Million Women so much, and so much of Tara is in everything we do.

In 2018, we launched the 1 Million Women Tara Fellowship. Each year, the fellowship will offer a young woman the opportunity to undertake training to expand her skills and knowledge in communicating climate action.

'This fellowship is not just about communication of the facts and stats but the challenge to connect people to the gloriousness and gift that our natural world is ... It's really about love because, as women, I believe that is our essential gift to life.'

— Tara Hunt, 24 May 2017

Get the 1 Million Women app!

To help you with taking action in your life we've been developing the 1 Million Women app. It has the best of what we've learned over the years about **how to empower people to change the way they live**. I'm talking bite-sized, daily actions that show you exactly what to do and the difference you're making – and exactly how many of us around the world are making those changes, in real time. It showcases the power of us when we all act together, globally and locally, and **it supports a live conversation feed so we can share our stories, our ideas and our solutions for climate action.**

The 1 Million Women App is free. It's an invaluable tool and I'll just wait right here while you download it ... Got it? Good!

The app will help you to profoundly change the way you live, and will help us all to better understand the challenges of changing our behaviour and lifestyles. We're also working with councils, cities, schools, universities, women's groups and many more organisations, everywhere. There's incredible power in acting locally and networking widely to create collaborations. If you are from any of these sectors or know someone who is, I'd love to hear from you. The app took two years to develop, with extraordinary support from a number of philanthropists and our community to raise the money to make it a reality. Over the next couple of years, we'll be taking the app around the world through a series of roadshows with the goal of enlisting 10,000 women to become 1 Million Women Ambassadors.

Want to be part of the movement?
The world needs everyday climate activists: that's our mission.
Love you to be part of it.

- Download our free 1 Million Women app – search for 1 Million
  Women in the Apple and Google Play stores (it's scheduled to
  launch near the end of 2018)
- Join up on our website and be counted – 1millionwomen.com.au
  And take part in any of our campaigns.
- Join our social media community – Facebook @1millionwomen;
  Instagram 1millionwomen; and Twitter @1millionwomen
- Become a 1 Million Women Ambassador. We need you.
  For more info, contact us at ambassador@1millionwomen.com.au
  You can access online slides and videos, and a guide for how
  to hold your own 1MW workshop with your friends, family or
  community, to give others the inspiration to profoundly change
  the way they live. It's all free.
- Send us an email at enquiries@1millionwomen.com.au

PART 1

SAVING
THE
PLANET

✓ A journey is a series of single actions, so **commit to starting right here, right now.**

1

Discovering my power

How it all began

Back in the day, I was a cosmetics manufacturer. I'd set up my own company in the 1980s after returning from a long stint in London, where I'd been inspired by the growing success of Anita Roddick's The Body Shop. I was young and naïve, but slowly I built my range and reach, eventually getting my products into beauty salons and department stores. I quickly learned that the cosmetics business is all about packaging – that how a product is labelled, wrapped and boxed counts for as much, if not more, than what is inside – to the companies making the decisions about what to stock on their shelves. Not that it bothered me particularly back then. It was just what I had to do to get my range into stores. I even used microbeads in a few of my products. And although I was married to an investigative and environmental journalist turned sustainability advocate, I wasn't particularly engaged with the issue of climate change or the environment.

Until I had an epiphany.

Now, epiphanies are fascinating, wonderful things. Mine was brought on by a series of seemingly everyday events. Who'd have thought it would change everything?

The year was 2006. My life was chugging along on its regular path. I was busy running my business and, together with my husband, Murray Hogarth, was immersed in the beautiful day-to-day world of raising our four children – Bronte, Jacob, Shea and Isaac – who were at varying stages of their school education.

Then the normal rhythm of life was disrupted. It wasn't just one big thing, but rather a bunch of seemingly minor distortions.

First, the unforgettable documentary *An Inconvenient Truth* came out, which I went to see with my daughter Shea, then 11 years old. The Oscar-winning film follows former US Vice President Al Gore during his campaign to raise awareness about human-induced global warming. It proved a turning point for many, in terms of how they viewed climate change, a topic that until then had seemed complex and hazy.

Shea and I came out disturbed, shocked and angry – the message of this film was impossible to ignore.

Sadly, in my own country, as brutal proof of Mr Gore's arguments about our planet getting hotter and the associated dangers of this, we were experiencing one of the worst early-onset bushfire seasons in our history. Australia's south-east was ravaged by fires that saw a million hectares burned over the course of two horrendous months. That was in the spring, with the scorching heat of an Australian summer still to come.

With a natural disaster taking place on our own soil, and the film's timely release, people were beginning to discuss the issue of climate change more openly than before. The media, too, was starting to get the point. **Suddenly, journalists and reporters were talking about climate change in language that the rest of us could understand. People – myself included – were starting to say, 'I get it.'**

Then, in the midst of all this, I was asked by a family friend and business owner to teach his team how to sell. Working as a cosmetics manufacturer in the hard-sell world of beauty, I was a pretty powerful salesperson, well versed in the ways of promoting and marketing skincare products. My friend's product was energy-efficient lightbulbs, which his company, one of Australia's pioneering greener lifestyle businesses, was offering to install in people's houses free of charge.

As selling points go, this product had plenty. These new energy-efficient lightbulbs were revolutionary. They could last up to 20 times longer than the older-style incandescent bulbs they replaced. They could save us money, cutting the size of our power bills. More importantly, they used up to 80 per cent less energy than regular bulbs, significantly reducing the amount of harmful carbon dioxide emissions released into the atmosphere. And the bulbs were *free* because the company was paid via a government scheme designed to help cut pollution contributing to global warming.

So I met with the team and started to teach them what I knew. If I could sell cosmetics, I figured giving away money-saving, energy-efficient lightbulbs was going to be easy.

At the same time, Murray, who has always been an environmental crusader, was writing a book about Australia's climate-change wars. He asked me to read through the manuscript and give him feedback from a layperson's point of view. He knew I was about as far from a climate expert as anyone on the planet, but experts weren't the readers he was writing for.

It wasn't that the subject of climate change was new to me. I was aware that there was something going on and, of course, with my new-found knowledge of energy-efficient lightbulbs, I was becoming better informed. But the topic had always seemed so overwhelming and so removed from my own life. At dinner-table discussions, when talk of climate change came up (which it inevitably did), I'd stay silent as I felt I had nothing to contribute. For me, it was always someone else's issue. After all, I was only one person in this complex global situation – what could I possibly do to help?

I wasn't ignorant or unaware, but I was disengaged, disconnected and disempowered. I can see now that the way I was back then, the powerless me, was a fundamental barrier to action. If you don't know enough about an issue, and if you can't see how you can make an impact, then it's so much easier to say and do nothing. That was me.

As I read the manuscript of Murray's book, I couldn't help but become more engaged. **'What do you mean, we could lose the Great Barrier Reef?'** I asked him. 'Are you saying this is what's going to happen if we don't change course?' I wanted to understand more and, as my knowledge deepened, I started to become more connected to the issue.

But that connection was still very much in my head. My own life, my own actions, continued along the same path as before. I was grasping the gravity of an emerging climate crisis, but I still didn't feel it in my heart.

✔ 'Getting' something isn't the same as truly understanding it. It isn't the same as feeling so deeply moved that you have to act.

The moment that changed everything

All these things were happening around me, but it took one final episode to bring me to my epiphany. One night, I was out with the lightbulb team, who were celebrating the fact that they had just installed a million bulbs in people's homes. That was a good reason to celebrate. It was one of the biggest mass consumer action campaigns in the history of environmentalism. I remember looking around at the 200-odd people in the room and thinking that the only person who hadn't contributed to this extraordinary achievement was me. Sure, I'd *talked* to them about selling, but what had I actually *done*?

It was then that the huge disconnect in my life – and in our house – became startlingly evident to me. There was my husband, literally going off to work to save the planet, and there was I, submerged in an industry dedicated to overpackaging. I didn't even have my head around separating the household recycling. There I'd been, helping this company install energy-efficient lightbulbs in people's homes, while we still had our energy-sucking halogen downlights blazing away. Our world at home, the dynamics of our family life, was completely unrelated to the world we were engaging with outside.

When I got home after the celebration, I looked around our house. 'What can I do right now?' I asked myself.

The issues of electricity and saving energy were on my mind, so I focused on that. I set out to find ways I could reduce the energy use in our house, something we'd never really considered before.

I started turning off appliances at the wall, rather than leaving them on stand-by. I looked at how many lights we'd leave on in our house, even when we weren't in the room, and I began turning them off.

The next time I received our household electricity bill, I couldn't believe my eyes: it was down by 20 per cent. Not only had I managed to save us a sizeable amount of money, but I had also cut down our quarterly carbon emissions by more than 250 kilograms. **The actions I had taken, as small and simple as they seemed at the time, had achieved a lot. I had achieved a lot.**

I can still remember the moment I saw the results of my actions, right in front of my eyes, and the feeling of exhilaration when I realised one truth.

I am powerful.

Action is empowering

The unique sense of power that filled me took the place of the helpless feeling I'd had, that 'There's nothing I can do.' Hadn't I just done this incredible thing? What else could I do if I made other changes to our lifestyle?

More importantly, that sense of empowerment propelled me to a position where I not only 'got it' in my head, but I also began to truly understand in my heart that I could play a vital role in the care of our planet.

That first small step to reduce our energy consumption propelled me from the camp of inaction to the camp of action. I was inspired by my achievement to do more – I went on to reduce our family food waste next and got it down by something like 80 per cent. Eventually, I ended up taking action in all sorts of areas, and my family came along, too. These included:

- Energy
- Food
- Plastic
- Overconsumption
- Fashion
- Money
- Transport

A single person can make a difference

What happens when you've had this incredible epiphany and your world has been completely altered? Well, in my case, you decide to start a women's movement.

When I took these life-changing steps, I'd never marched in the streets for any cause, and I'd certainly never started a social movement before. I wasn't a green role model, and I've never been an environmentalist.

I just had this burning belief that, if I could get my story out to other women, they would act in their lives like I had in mine. I felt there must be so many women out there just like me, baffled by all the jargon, intimidated by the horror stories and not knowing what they could do.

A few years later, 1 Million Women was born with the tagline: 'We are daughters, mothers, sisters, grandmothers inspiring climate action'. It started with one woman waking up, and it is now almost 10 years old and 800,000 women strong (and counting). In the past 12 months, 4.5 million women around the world have read our blog, and every week we reach around 10 million women through our social media channels. We have received an award from the United Nations Momentum for Change: Women for Results recognising the work we do. Every day, women share their stories and challenges, and offer encouragement and advice. Every day, we are making our mark in the fight against climate change through the way we live.

And every day counts.

When I got the point about climate change – when I truly understood it in my heart as well as my head – **I realised that it is our love for this planet and our concern for its health that are going to drive us to fight for it.** Now is the time to take our share of responsibility for this beautiful planet – we are its caretakers, and it needs all we can give.

In the chapters that follow, I'll tell you a whole lot more about what you can do in every aspect of your life to reduce your impact on the planet. You'll be amazed by how easy and empowering it is, and how you can change your life forever.

Before we find out everything we can do, we need to face the facts. The way we live our lives is contributing to overconsumption, waste and pollution, which, combined, is now threatening our planet. The effects are overpowering nature itself; the changes to Earth's climate are one of the most alarming symptoms.

Yes, together we can change things, but we need to know what we're up against first.

❶ Climate change is not waiting for us to wake up and catch up.

Carbon pollution is rising; we are spending more and consuming more, filling our oceans with plastic and our atmosphere with damaging gases.

2

What's the problem exactly, and what can we do?

Truth bomb time

Twenty years ago, my husband, Murray, wrote a cover article for *The Sydney Morning Herald's Good Weekend* magazine, titled 'All I Want for Christmas is a GREAT Barrier Reef'. Having grown up on the Queensland coast, Murray had a deep personal connection to the reef. In the story, he tracked the decline of a truly great natural wonder, lamenting how efforts to manage the World Heritage marine treasure better were visibly failing, especially as the existential threat posed by climate change was becoming more apparent.

The threats to the reef that Murray wrote about back then – heat-driven coral bleaching, industrial coal and shale oil mining, port developments, and nutrient and chemical pollution impacts from land-based farming and cattle grazing – are sadly even more prominent today.

With the article, Murray was the first Australian journalist to identify the fresh major threat (now far better understood in scientific terms) of **ocean acidification caused by higher levels of carbon dioxide in the atmosphere** (also driving climate change) being absorbed into the oceans, turning them more acidic.

That's chemically very bad for corals and shellfish, inhibiting the ability of many marine creatures to form exoskeletons – the very building blocks of a coral reef.

! The past two years have seen the Great Barrier Reef suffer **increasingly more severe bleaching events**, which not only kill more corals, but also mean that any recovery will be harder.

The size of about 70 million football fields, the Great Barrier Reef is the largest living structure on our planet. There's now no question that it and coral reefs everywhere are in peril because of human activity, our industries and the things we do in our daily lives.

Murray was flown over the reef in a light aircraft by alarmed scientists, who pointed out the ghostly white, dead-coral legacy of one of the first waves of so-called 'bleaching events', which have since become increasingly commonplace.

According to the Great Barrier Reef Foundation, **between 1985 and 2012, human influences have resulted in a 50 per cent decline in coral cover.**

I think back to Murray's article all those years ago and look at a recent report from the Great Barrier Reef Marine Park Authority, the official guardians of this World Heritage wonder, which says the greatest risks 'remain unchanged'. So much time has passed between these two publications, yet we've done nowhere near enough to save the reef. Time is running out fast. This precious treasure could be wiped out in our lifetime.

In order to talk about climate change and what we can do to fight it, we need to get the full picture. I want to give you a snapshot of the current situation. I can't even promise that what I'm writing now will still be current when you read it, although, something tells me the picture won't be any brighter.

So let's do it now. I need to take you to the dark place, the honest reality of what's happening with Earth. I won't leave you there for too long, but let's do it now, then identify the steps we can take to move up from there together.

> ❗ **People under 33 have never experienced a month that was cooler than average.**
>
> The last time the global monthly temperature was below average was 1985.

The hard facts of climate change

Every day, we see alarming news about the destruction of the environment: plastic clogging up our oceans, our precious species becoming endangered or going extinct, lethal fires blazing out of control, extreme weather events and record-breaking temperatures. Sadly, it's all true. Our planet is under threat, and it's not some natural evolution.

Ninety-seven per cent of the world's scientists have reached a consensus that climate change is real, it's dangerous, and the evidence overwhelmingly shows that it is caused by humans.

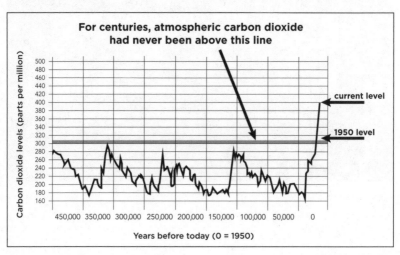

climate.nasa.gov

What is climate change?

Climate change literally means that climate patterns are changing, regionally and globally, including the seasons. Primarily, this is a consequence of global warming, which is being driven by the effect of rising levels of carbon dioxide and other 'greenhouse gases' in Earth's atmosphere. This is also known as the 'greenhouse effect', because the atmosphere is warming like the air inside a glass greenhouse for growing fruit. And there's no question that carbon dioxide levels have risen – scientists have shown that current levels of carbon dioxide are higher than they've been at any time in the past 400,000 years. Now, for the first time in human history, carbon dioxide levels are higher than 400 parts per million (ppm) and, if we continue to burn fossil fuels such as coal, oil and gas at current rates, while clearing the forests that breathe in carbon dioxide, the levels could reach as high as 1500 ppm.

How did we get here?

Relying on fossil fuels

Since the birth of the Industrial Revolution in the 18th century, there has been a relentless increase in carbon dioxide pollution in the atmosphere. In fact, almost as soon as fossil fuels were introduced – first with coal-fired steam engines, then big new factories that replaced small cottage-style ones,

> ❗ In just 200 years – a geological blink of an eye – we've cut down vast areas of forest, **unlocking huge buried resources of fossil fuels from Earth.**

and later the rise of industrial-scale agriculture – the levels of carbon dioxide in the atmosphere have been rising.

All that fossil fuel contains carbon that had been taken out of the atmosphere hundreds of millions of years ago and safely stored underground, as a natural process. When it's burned, that stored carbon dioxide is released. At the same time, human practices such as industrial-scale manufacturing, farming and land clearing have contributed to the problem and caused further environmental damage.

Losing touch with nature
To make matters worse, many of us have become disconnected from the natural environment that has nurtured humanity for so long. We've lost touch with nature and, in the wealthier parts of the world, we've become locked into waste-filled, polluting,

overconsuming lifestyles that harm the environment that
sustains life on Earth.

Population growth

A big part in this story has been played by population growth.
It wasn't until the Industrial Revolution that, for the first time
in our species' existence, the world's population hit 1 billion
people. Now, just over a mere 200 years later, the number of
Homo sapiens is almost 8 billion. So sheer population pressure
is another part of the equation.

Ignoring the warning signs

Unfortunately, we mostly missed the early warning signs that
our rapid growth was heading out of control. And when the
dangers became too obvious to overlook, many downplayed
them. Maybe you've known a person like humanity: they had a
lucky run, became arrogant and increasingly reckless, then it all
started to unravel. That's us, circa 2018.

The trouble we're in isn't anything new – it's been building
for more than two centuries. But it has accelerated dramatically
in the past 50 years, and now we've reached five minutes to
midnight.

Then there's the problem with plastic ...

Plastic is one of today's headline pollution problems. In fact, the plastic problem is very much tied up with climate change. For one thing, **most plastics are made from oil or gas, fossil fuels that require the use of even more fossil fuels for their extraction and production.**

Take 'disposable' plastic water bottles, for example. The carbon footprint of one such bottle is 82.8 grams of carbon dioxide. It doesn't sound like much, until you consider that we throw away 563 billion single-use plastic water bottles every year. This means that, **every year, the production, transportation and disposal of plastic water bottles produces 46 million tonnes of carbon dioxide.**

Then there's the problem that plastic is virtually indestructible. This, combined with its relatively low cost, means that we not only use and throw away far too much of it, but also that it never goes 'away'. Instead, it is now clogging up our rivers and oceans, and posing a serious threat to marine life, birds and animals. In fact, it's been predicted that, by 2050 (just three decades away), there will be more plastic in the ocean than fish.

The proliferation of plastic is a major problem we're facing. Rather than ignore it, we need to take action now.

Quick ways to cut back on plastic

DO TODAY:

- [ ] When someone offers you plastic, just say No!
- [ ] Remember to use your reusable bags!!!
- [ ] Take your own containers to be filled at the takeaway, deli or bulk wholefoods shop.
- [ ] Leave overpackaged goods on the shelf.
- [ ] Ditch the straws (unless you have to use one).
- [ ] If you forget your keep-cup, drink your coffee at the café.
- [ ] Don't use coffee pods.
- [ ] Stop using plastic wrap.
- [ ] Reuse the plastic you already have.
- [ ] See if you have a sturdy bottle at home to use instead of a plastic water bottle.
- [ ] Properly recycle any plastic that can't possibly be reused.

PLAN TO DO:

- [ ] Switch to stainless-steel water bottles.
- [ ] Turn an old T-shirt into a tote bag.
- [ ] Gather a collection of reusable bags to share with friends and co-workers.
- [ ] Make your own beeswax wraps as an alternative to cling film.
- [ ] Seek out companies doing all they can to eliminate disposable plastic. They deserve our support.

> **!** Every year, **8 million tonnes of plastic** is being dumped into the ocean. That's the equivalent of **one garbage truck dumping its contents into the sea every minute.**

What are the effects of climate change?

Effects of the climate-change phenomenon are already all around us, from rising sea levels to more intense weather-related natural disasters.

Extreme weather

Higher average temperatures, in the atmosphere and also in the oceans, are the most obvious consequences of the global-warming trend, which in turn is the engine room of climate change. On average, global temperatures have risen by 1°C during the past century. Many climate scientists believe that, by the end of this century, temperatures could increase by 5°C, a rise that could be catastrophic for the planet and its inhabitants.

These rising temperatures mean even hotter extreme heat waves, more powerful storms, including tropical cyclones with higher destructive wind speeds, and greater volumes of torrential rain, leading to soaring flood dangers. The more severe floods and storms, which used to be classed as one-in-a-hundred-year events to highlight their rarity, are becoming more frequent.

We're already noticing the effects of global warming in our own lives. At the start of 2018, Sydney had its hottest day in 80 years. The year before, every state in the USA had a warmer-than-average year, with 32 states recording one of their ten hottest years on record.

Globally, 2014, 2015 and 2016 were the three hottest years since 1850, when records began.

Warming also causes more severe droughts, as higher temperatures accelerate water evaporation. Terrifyingly, this means bigger and more lethal bushfires or wildfires, with longer fire-danger seasons across expanding areas – which is something that many people in Australia, the USA, Spain, Portugal, Greece and elsewhere are experiencing already.

Melting ice caps and rising sea levels
Global warming will cause the melting of the polar ice caps and glaciers, which translates into major sea-level rises; and also the

melting of permafrost areas such as the Siberian tundra, with the release of trapped carbon dioxide and methane from the previously frozen, now rotting, vegetation. Scientists predict this might take 1000 years or more, but the process is already underway.

How will it affect us?

On the human side, the adverse effects of climate change will impact most upon the people who already have little or nothing: those who exist in the most precarious circumstances and personally have contributed the least to the pollution of the atmosphere, because they live with few resources and can't afford to waste anything.

More than 1.5 billion people on Earth don't yet have electricity, and often live in countries where women still walk large distances every day to carry water and fetch firewood. **There's abundant evidence that weather-related natural disasters, such as floods and famines, cause harm to women and children in the world's poorest nations and, in a climate-change world, these threats get worse.**

I've been to tiny Pacific island nations and met with leaders from these micro countries, on the frontline to be swamped by rising sea levels. In some cases, the ocean is already intruding and people are contemplating abandoning their ancestral lands. Inevitably, they will be looking to nations such as Australia and the USA to take them in.

Climate change is personal, and we need to act now

These are the facts and, while climate change is global, at the same time it's personal and local. It's the energy I use, the food I eat, the clothes I wear, the car I drive and the waste I create. **It's not something that will only happen in our children's children's lifetime – it's happening now, and it's accelerating.** It's now part of all our stories, a set of huge challenges that we're all responsible for solving.

Nor is it a short-term problem. Some of the carbon pollution currently sitting in our atmosphere went up there in the 1970s; what we're pumping out today will be there for the next 30 years. Other greenhouse gas pollutants, such as methane from rotting organic waste, are even more long-lasting.

In fact, climate change is happening faster than predicted by scientific models only two decades ago. Seasons are visibly changing, crops are being disrupted, reefs are dying, species are on the move and mass extinctions loom.

17% of australian households have solar panels on their roofs. **That's almost 1 in 5 households generating their own solar energy.**

Australia is leading the solar rooftop revolution – people are taking power into their own hands.

The only real chance we have of avoiding the dangers of runaway climate chaos is to _act now_.

Climate change can't be pushed aside in our own lives as something we can leave for future generations to worry about.

Doing nothing isn't an option.

But what can we do?

What we can do

When I say 'we', I'm not putting the onus all on myself, or on you, either. This is bigger than all of us. Our response to climate change has to be multifaceted. **We need all levels of government to act, as well as industries and corporations, large and small businesses, communities, households and individuals.**

What binds any society together is its people, and we all need to get involved. It's got to be a top-down and bottom-up thing.

Every single one of us has our part to play, both individually and together.

So where do we start?

Take action – it's the antidote to being overwhelmed

Confronted by the facts about climate change, and the fear of what's to come, it's easy to feel hopeless and helpless. At one end is denial: 'I don't know what's happening. I can't see anything different, and I don't really believe all I'm hearing, anyway.' At the other is despair: 'This is all too big for me. I'm just going to keep living my life – I can't afford to worry about what may or may not be going on around me.'

But between denial and despair, there is a glorious bit in the middle: action.

And this is where we come in. One individual's action *can* have an effect.

Every movement, every revolution, every sweeping global change starts with the individual. We are all a big, important part of this picture. And we must never underestimate the power we have in it.

If enough of us began living differently, then our families, communities, cities and countries would reflect that difference. This is how systems change.

To live differently, we need to change our behaviour – and that's in our own control. If we opted for less packaging, there would be less harmful plastic clogging our oceans. If we turned off our appliances when we don't need them, our households would release fewer carbon emissions into the atmosphere. If we were more mindful of the purchases we make and things we discard, less waste would end up in landfill. Our choices can literally make a world of difference.

Then, we need to connect with our community, to share our efforts and ideas, lobby our politicians, use our vote, call out corporations that aren't acting, get the kids involved and bring other people with us.

But before we go deeper into exploring how we might change our lifestyles, reconnect with the planet and mobilise our communities, I want to address a really important issue. It's one that may have been nagging you since you picked up this book.

Why women? Why put this on us?

The solutions to global warming and climate change cannot fall on women's shoulders alone.

This needs us all, women and men.

But women are extraordinary change makers, powerful networkers and, well, it's pretty simple when you've seen what I've seen.

Quick ways to cut home energy use

DO TODAY:

- ☐ Switch off lights.
- ☐ Turn off electronics and appliances at the wall when not in use.
- ☐ Wash clothes in cold water.
- ☐ Hang washing out to dry.
- ☐ Clean fridge and dryer filters to improve their efficiency.
- ☐ If you have a second fridge, switch it off when not in use.
- ☐ Take shorter showers, under four minutes.
- ☐ Close off rooms and areas when not in use.
- ☐ If you have a pool, set your pool pump to run an hour less.
- ☐ Set your heating a few degrees cooler.
- ☐ Set your cooling a few degrees warmer.

PLAN TO DO:

- ☐ Install energy-efficient lights.
- ☐ Choose energy-efficient appliances.
- ☐ Install a low-flow showerhead.
- ☐ Install a water-efficient toilet.
- ☐ Seal draughts and gaps.
- ☐ Install thick, close-fitting curtains.
- ☐ Install ceiling insulation.
- ☐ Replace your air conditioner with a fan.
- ☐ Replace your gas or electric hot-water system with solar-powered.
- ☐ Recycle your second fridge.
- ☐ Install rooftop solar panels.
- ☐ Investigate going off-grid with rooftop solar and battery storage.

3

Why women?

Because we're powerful!

'You are, you are courageous
Walk on, walk on
These are the times
that can make us
courageous and strong'

These stirring words are from the song 'Courageous' by Australian country music legend Melinda Schneider. The multi-award-winning singer–songwriter gave this song to 1 Million Women in our early days to use as our anthem.

Gratitude doesn't even begin to cover how I feel about Melinda's beautiful gesture. In her hectic life of performing, I cannot think of a time when she was too busy to sing at a 1 Million Women event. What a wonderful ambassador she's been.

At first, at the start of my 1 Million Women journey, I was operating mainly on instinct. The idea that a mass movement of women could unite to act on climate change in our daily lives just felt right to me. After all, hadn't I just saved 20 per cent on my family's energy bill?

If one woman could make a difference by taking a few small actions, imagine how powerful millions of us would be!

To build on instinct, I needed evidence. Early on, it became clear to me that when it comes to the well-being of the planet, men and women are not equal. For starters, mainstream market research showed that, on average, women were about 10 per cent more concerned about the environment than men. Over the years, I've seen more and more such research, and the results are consistent.

Then I learned something even more compelling, something that I think I already knew in my gut to be true. Depending on whose numbers you use, in high-consuming economies, such as Australia, the USA or the UK, **women make between 70 and 85 per cent of all the purchasing decisions that affect a household's carbon footprint.** For me, that was a real 'I get it' moment. In a consumer-led society, making the spending decisions is power, and brands know it.

Now, in case you're wondering, I am not anti-men. I promise!

1 Million Women is not about being anti-bloke. Rather, it's about amplifying the strengths of women, mobilising our power as change makers and using that to shape the sort of world we want to live in.

Women control a lot of money, and the trend is up!

It's estimated that women now control $29.6 trillion of the world's wealth. That's just a little more than 10 per cent of total global wealth (at around $280 trillion) but it's trending up and, by the time you read this, the statistic for women will most likely be more than $30 trillion. And when you take our spending power into account, we influence a lot more dollars.

In fact, by 2028, women will be responsible for about two-thirds of consumer spending worldwide. Think about all those major spend-up-big events that shops rely on: Christmas, Valentine's Day, Easter, Mother's Day, Father's Day … In the majority of households, it's likely that a woman made the key spending decisions. Cha-ching!

In other words, women are the biggest spenders on the planet. We're cash machines that brands and businesses want to tap into for all we're worth. No wonder advertisers love us so much!

So, apart from the depressing fact that we're the favourites of advertisers, what does that mean? It means that we're *powerful*. That we're a force to be reckoned with. And that, with every single dollar we spend, we can shape the kind of world we want to live in.

Our spending (or non-spending) choices have a major impact: they can make or break businesses, change store policies, drive consumer trends – and help save the world.

Our financial power doesn't end at the check-out, whether in-store or online. With women controlling more and more of the world's wealth, how and where we invest our assets becomes more powerful as well. If we don't like a bank continuing to invest in coalmines, for example, we can shift our funds and tell the bank why. The same goes for our superannuation or pension funds, and any share-market investments we make. Ethical and sustainable investment options are becoming more widely available every year, and they often rank among the better financial performers. Every time we make an investment choice to protect the environment, we send a message to the big end of town that we care, and that we have the financial clout to do something about it.

> 'One woman can do a lot when you look at one woman with another million women who stand alongside her.'
>
> – Rachel Perkins, Australian film and television director and 1MW Ambassador.

Women are great networkers and agents for change

Who better to come together in a movement than women? In my experience, women are natural networkers who will work intuitively to get practical outcomes in almost any situation. If things aren't right, you'll always find women stepping up to be agents for change for the better.

> 'If one woman in every family takes the lead on this, the world will change.'
>
> – Wendy McCarthy AO, 1MW Ambassador

✔ **Why is political representation important? Women can make a difference.** In India, research showed that the number of drinking-water projects approved by *panchayats* (local councils) **led by women was 62% higher** than in male-dominated councils.

When it comes to climate change, women get a bum deal

When I first started thinking about the idea of a women's movement, I knew that women face a gender pay gap in most countries, including my own, and that equality is still wanting in many areas. What I hadn't realised until I started on my path with 1 Million Women was that, when it comes to climate change, women get a really bum deal.

In the developing world, it is women and children who will suffer most from the impacts of dangerous climate change, just as they suffer most in weather- and climate-related natural disasters, such as floods and famines.

According to the United Nations, 80 per cent of those displaced by climate change are, or will be, women and children. This is because, in developing nations, women are on the frontline as the primary caregivers and providers of food, water and fuel. They are the most vulnerable, sometimes for simple, but tragically practical, reasons such as not being able to swim, or the brutal reality of having to walk long distances daily to collect water and firewood. In the aftermath of natural disasters, they are more likely to suffer from poverty, disease, loss of housing and income, and violence from men.

> ✔ New Zealand Prime Minister
> Jacinda Ardern has set a goal
> for the country to plant
> **100 million trees** each year
> to combat climate change.

As a woman living in one of the world's most affluent societies, my experiences are obviously very different to those of women in developing nations. Despite that difference, or perhaps *because* of it, a realisation hit me once I started to understand it more deeply. **I have a responsibility to do everything in my power to support women on the frontline of climate change.** Shouldn't I do all I can to live with the least impact on the planet? After all, we live under and breathe from the same atmosphere.

Just how women will be affected by climate change was driven home to me when I attended the Micronesian Women's Conference in the beautiful, but threatened, Marshall Islands. This small island state is located deep in the vastness of the Pacific Ocean. Flying there from Australia, the plane island-hopped through a series of tiny Pacific nations, many of them low-lying and in the first wave of countries whose existence is threatened by rising sea levels and more powerful tropical storms. Seeing the impacts of climate change firsthand, with water lapping the land, opened my eyes to the fight the Marshallese and all small island nations in the Pacific are facing. It dug deep into my heart.

Her Excellency, President Hilda C. Heine of the Marshall Islands was a gracious host and generous with her time to all conference goers and to me personally. She is the eighth president and first woman to hold the position, and she told me that, right now, a third of the country's population, including women who rely on the production of traditional handicrafts, are being affected by climate change.

This wasn't some theoretical issue that might or might not happen in the distant future. This was happening to women, right there, right now. As Her Excellency said:

> 'It's important to integrate and get women involved as much as possible in the government of their countries … Climate change is impacting our communities on all different levels because we've experienced droughts last year and again this year. And the effects of those droughts on women have been extraordinary. As you well know, **women are the managers of their home; they are responsible to make sure there is enough water for drinking, for cleaning the clothes.**

And these women don't have enough water. They have
to walk long distances. It's impacting their lives. And the
saltwater intrusion impacts our food plains and many of our
crops like our taro and fruit.'

Vonda's island home

In Australia, Vonda Malone was elected the first female mayor of
the Torres Shire Council in 2016. The Torres Strait Islands lie at the
north-eastern tip of the Australian continent. With her island home
already feeling the threat of inundation from the sea and the ravages
of climate change, Vonda is committed to raising the voices of the
Torres Strait Islanders, especially the voices of women. **'I would
hope that I could mentor and support other women to consider
putting themselves forward, so I won't be the lone voice at the
table.'**

'Incorporating women in solving climate
challenges is an imperative – we make
up half of the population and represent
3.5 billion ways to change the world.'

– Lorena Aguilar, Vice Minister for Foreign Affairs
for Costa Rica, and my dear friend! x

We need to shout louder in the climate debate

While women have the greatest consumer spending power, are the people most adversely affected by climate change, and make up more than half of the world's population, the fact is our voices are currently underrepresented in the arena of climate change when it comes to positions of power. United Nations figures show that women hold only 23 per cent of legislative posts. It's crazy to think that the people most affected by climate change have less opportunity to contribute to the most important discussions about its solutions. The real point is that climate change is far too important to be left to men on their own, or to women alone for that matter. **We're all vulnerable under one atmosphere, and we all need to be part of finding lasting solutions.**

It's not all bad news, however. Since starting 1 Million Women, I've witnessed and been part of significant and extraordinary change. Now, when it comes to climate action, more and more women around the world are leading the way, turning ideas into solutions.

Just look at the recent breakthroughs in policymaking. The 2015 Paris Climate Change Agreement – the historic global plan signed by 195 countries to cap temperature increases – noted that **'climate change is a common concern of humankind'** and that,

Women mayors across the globe are leading on climate action

At the inaugural Women4Climate Conference, in New York in 2017, **15 women mayors from around the world pledged to reduce CO$_2$ emissions and fight climate change.**

Responsible for more than $4 trillion dollars in GDP and governing 100 million urban citizens in some of the most influential C40 cities, including Paris, Mexico City, Cape Town, Caracas and Washington DC, the women mayors were joined by business leaders and influential change makers to promote climate action and champion the next generation of female leaders.

'As women, we know all too well that the powerful often seek to silence our voices when we speak out to protect the most vulnerable in our communities,' said C40 Chair and Paris Mayor Anne Hidalgo. 'We are here today to show that we refuse to be silenced. All around the world, in city halls, corporate boardrooms and on the streets of our cities, **women are demanding action to protect the planet from the threat of climate change.**'

when addressing climate change, parties should consider 'gender equality, empowerment of women and intergenerational equity'.

Good, right? Want more? At the UN Climate Change Conference in Bonn in 2017, governments adopted the first-ever Gender Action Plan, which aims to increase the participation of women in all UN climate-change processes.

'Women are at the heart of dealing with climate change, building resilience in communities as climate becomes more unpredictable and disruptive. **Climate change needs to become a key issue of the global women's movement**, and women leaders should give it much more priority at every level.'

– Mary Robinson, former first female president of Ireland and head of the Mary Robinson Foundation – Climate Justice, and great friend of 1 Million Women

My friend Bridget Burns, co-director of a global women's advocacy organisation called WEDO (Women's Environment & Development Organization), has fought for progress on this issue every step of the way. Bridget, and the many extraordinary women who stand alongside her, has worked tirelessly on behalf of women across the planet. She sums it up this way:

> '**Gender equality is now more universally recognised as a vital, cross-cutting issue to be considered in developing climate policy** ... I have been privileged in my decade of work on climate change, not just to see changes in policy, but to witness the strength in women building power across countries and movements, to demand together a more sustainable future. This is where my hope comes from.'

While we will always want things to move faster, progress is being made.

When I asked my dear friend Christiana Figueres, who led the United Nations climate team that delivered the historic Paris Agreement in 2015, to share her insights, she said:

> 'We often hear that women are needed at the forefront of climate change because women around the world are disproportionately negatively affected by climate change. This is very sadly true. But in addition, what I think is also

urgently needed is women and men who are conscious of their feminine intelligence and energy. Women and men who understand that to protect the long-term we need to act in the short-term; **women and men who know that we are on this planet as stewards and improvers, not as destructors**; women and men who intentionally support each other and collaborate in the full wisdom that only through radical collaboration can we effectively deal with the complexities we face ...'

At 1 Million Women, we are all part of a global movement of women changing the world. There is one story, and we are in it. Yes, we're all facing different degrees of challenges, including women on the climate-change frontline fighting for their land and the survival of their way of life. But we're in this together, and we can show that by changing the way we live every day.

Solidarity doesn't always have to be spoken aloud. The support we show through our actions connects us to every woman in the world.

So if someone asks me 'why women?', the answer's a no-brainer: because women are most adversely affected by climate change and women have enormous power to do something about it.

I know how easy it is to do nothing if you think you don't know enough. To mentally tick the 'not sure' box. One of the earliest messages of 1 Million Women addressed this. It is just as relevant when I write it again today:

Try not to become paralysed through worry about all the jargon, what the technical words mean and what is the latest policy on climate change. Forget about what you don't know.

Just do one little thing – something for which you can see a result. I did and, I promise you, it will lead you to something else. Then repeat.

The rest of it – the awareness and detailed knowledge, the curiosity and hunger to know even more, and the determination to do more – will come along the way. This is the action that moves you forward on this journey. It's the action that empowers you to do the next action. I know this to be true because that's exactly what happened to me.

Never underestimate the power you have as one woman, one person to change the world.

Christiana's four-point plan

Christiana Figueres, who led the United Nations climate team for the Paris Agreement, puts increasing equality for women at the core of solving the climate crisis:

1. Reduce consumption patterns that have become incompatible with a sustainable future and, instead, **invest in products that support low-carbon living.**

2. Support the expansion of women's rights throughout the world, as well as their leadership in climate-related activities.

3. Enable the transfer of technologies to developing countries to help them establish renewable energy and build sustainable transportation. This includes technologies that will empower women to adapt to climate change.

4. Encourage government representatives to achieve an international agreement on climate change, backed up by national plans of action. This will have a positive and lasting effect for all people.

The coolest chicks on the planet

Through 1 Million Women, I have been privileged to meet so many women who bring their unique knowledge and talent to the table of climate action. Some are now my dearest friends, and all have shaped the way I think. I have already mentioned a few in this chapter, but there are so many. Women such as:

Farhana Yamin, the environmental lawyer and policy expert who almost single-handedly created a compelling plan for a net zero carbon world economy by 2050. She sums it up this way:

> 'Representation and participation are fundamental human rights. Women, and men, must be equally involved at every level of decision-making, whether it is in their own families, their communities, their workplaces or at grassroots, national and international global-policy arenas. Taking up positions of leadership in politics, the media, the world of business and community groups is vital if we are to cater for all of humanity, not just one half.'

Rachel Kyte, CEO of Sustainable Energy for All (SEforALL) and Special Representative of the United Nations Secretary-General, focuses on providing access to affordable, reliable, clean energy and cooking to the one billion who don't have electricity and the three billion who don't have access to clean cooking.

Liane Schalatek, who makes climate finance more accessibly for the poorest countries – those in Africa and small island states – and those marginalised people, including women, already most severely impacted by climate change, in particular through her work on the Green Climate Fund (GCF) to ensure that its investments for climate projects also support human rights and gender equality.

Osprey Orielle Lake, founder of the Women's Earth and Climate Action Network (WECAN), who works tirelessly for women's leadership in climate solutions and the rights of indigenous women in this climate fight.

Sally Hunter, born and bred in rural Australia, who is driven to protect rural landscapes and precious water resources from invasive coal and gas fields. She helps her community transition away from fossil fuels and towards regenerative agriculture and renewable energy industries through her local group People for the Plains and through the Lock the Gate Alliance.

Tara Shine, who co-founded Plastic Free Kinsale to empower the people of her town Kinsale, Ireland, to reduce plastic waste.

Kathy Dede Neien Jetnil-Kijiner, the poet and activist from the Marshall Islands. She co-founded Jo-Jikum, meaning 'your home', a non-profit organisation to educate youth on environmental issues and to foster a sense of responsibility and love for the islands.

Noelene Nabulivou, from Fiji, who harnesses the strength of women from the small Pacific Island states facing wipe-out by rising sea levels.

Neha Misra, from Solar Sister, who helps eradicate poverty by empowering women in rural Africa through solar lantern projects and economic opportunities. Since 2010, 3200 women have provided 1.3 million people with transformational access to solar lights, mobile connectivity and clean cooking solutions.

And to the millions and millions of women across this planet who are taking action in their own local communities and households. They are changing the system through changing the way they live and demanding better from those who aren't.

'Ngaya yaanji yidaagay
MiiMi Wajaarr umbala.'

(I walk always with Mother Earth.)

– Aunty Bea Ballangarry, poet, author,
Aboriginal elder and 1MW Ambassador

4

Reconnecting with the planet

The number one reason to get our hands dirty

If there's one thing I know that will strengthen your determination to take lasting action on climate change, it's the beautiful world around you.

Walk in a forest, swim in the ocean, climb a mountain, stroll in the park and smell the roses (literally) – you'll be reminded of just how beautiful and important Earth is.

These days, we're all so busy and distracted, we're losing our connection with the only planet we have.

I truly believe that we need to rediscover a deep love for Earth if we are to drive the changes required to make a genuine difference. Loving Earth will underpin everything we do, as we go on this lifestyle-changing journey.

> ✅ **The hole in the ozone layer is now the smallest it's been since 1988.** Researchers say, 'A recovering ozone layer provides an example of the changes that are possible if humans take action.'

What's our relationship with our planet?

When I think about my relationship with the planet, I'm reminded of two family getaways. Both trips involved getting back to nature, but they couldn't have been more different.

The first was a holiday in northern New South Wales, Australia, to a place we often go to reconnect with nature. It's somewhere to walk in World Heritage–listed rainforests, explore beautiful creeks and waterfalls, and take in the magnificent open space and serenity. We love going there – it's a time to relax and regroup. And it's an important anchor point for us as a family to just stop.

This particular year had been a challenging one. But, let me tell you, Earth gave us everything we needed. We lay on the grass, melting into its gentle caress; we walked into a waterfall and swam beneath its cleansing flow; we watched the stars lighting up the clear night sky; we listened to the sounds of the rainforest; we found the space to breathe. The sweetest smells, the softest soil, the abundant design – nature's design – of the trees around us … every aspect soothed our spirits.

It was such an exquisitely pure, generous experience. Earth gave us so much for nothing. It seemed to embrace us and say, 'just be'.

A few weeks later, some of the family went camping at a farm in the state's Central Tablelands. We arrived to find a large swathe of the land completely denuded of life. The owners had cleared the area of native vegetation because they wanted to grow a certain type of fruit tree there. However, it turned out that these trees were not suited to the area and had struggled to thrive. Many had died. There had been real environmental loss for no economic gain.

I remember standing in the middle of this barren, dry landscape, feeling devastated at the death surrounding me. This piece of Earth looked as if it had been torn open. I simply could not see how it could ever be brought back to life. It was deeply unforgiving.

And the damage had been done at the hands of humans. It was humans who had cleared the land, wrenched out its life force for a perceived economic opportunity, which had failed anyway. In this case, Earth did not look like it had enough power to replenish itself.

The contrast between that first trip, where Earth nourished and replenished us, and this one, which showed how our species can use and abuse it, couldn't have been more stark. It was shocking, tangible evidence that we cannot keep taking from Earth without giving back. And we certainly can't expect Earth to just keep giving boundlessly.

Love the planet like family

The year after my epiphany, during my first thoughts of wanting to start a women's climate-action movement, I was privileged to be trained by Al Gore. On the back of his groundbreaking documentary, *An Inconvenient Truth*, Mr Gore was touring the world, including my own country of Australia, to increase awareness of climate change at a grassroots level and to train volunteers to present a slide-show version of his film.

I eagerly signed up for what is now known as The Climate Reality Project and, together with 70 others, spent three inspiring days exchanging ideas, discussing realities and searching for solutions. Among many unforgettable moments, one continues to resonate with me.

Al Gore talked to us about the day he nearly lost his son, an awful event that is examined in the film. One moment he was holding his six-year-old son's hand, the next, his son had run onto the road and was hit by a car. In the film, Gore says that the accident altered his world view forever, leading him to consider the reality that what we take for granted about Earth may not be there for our children. Talking to us in person, describing how he sat by his son's side in the hospital, not knowing whether his son was going to live or die, Gore said that the feeling ripped him open to the very core. And then he told us: 'That's how I feel about climate change.'

We need to come to a point where we are compelled to take responsibility for the care of our planet.

Speaking for myself, for too long I'd had the knowledge but hadn't followed through to make real-life changes.

But empowerment followed action. As I took a few small steps – cutting our electricity consumption, reducing our food waste – I began to see the connection between the way I live and the planet I live on. I realised that everything I did was connected to Earth; its well-being was not someone else's issue. It was mine, and that of my family and friends. Everything we do makes a difference.

After all, fighting for its health is fighting for the health of humanity.

We need to discover that connection and engagement with Earth, and come to love it in the same way that we love our family and loved ones. Because the truth is that we fight for what we love. We make sacrifices for what we love. That's why love is the most powerful weapon we have in our arsenal.

If you've ever fallen madly, deeply in love with someone, you'll know what I mean. You'll drop everything for them, trekking across cities, if needs be, to see them. If you have kids, you know that feeling, too. You'll sacrifice sleep, friendships, sex, even a little sanity for them; you put them first, 24/7.

We all 'love' going to the beach or for a bushwalk, but how do we actually fall in love with our Earth so we love it like our family?

This love needs to be a palpable force that can't be denied. It must be passionate.

Because, honestly, that's the place we all have to get to, to make our good intentions travel from our brains to our hearts to our deeds.

> ❗ Up to **40% of fresh produce is rejected by supermarkets** because it doesn't meet their cosmetic standards.

Reconnect with nature

Let me take you away for a moment. You're cycling along a narrow path, the incline steadily rising as you ascend a mountain. The trees on either side provide dappled shade along the path and, through the pretty, pale trunks on your left, you can see the crisp blue of the ocean. The sun is shining, the leaves are manifold shades of green, and you can almost smell their dewy scent.

Only you can't.

Because you're in a gym, and the natural world is on the screen in front of you. Sound familiar?

So much of our lives today is lived through the virtual, once-removed filter of technology. We sit captive to our phones, chained to our computers, hooked on our televisions and devices, which provide us with outlets for communication, entertainment and relaxation. **Everything is accessible, at our fingertips and on a screen before our eyes. No wonder it's easy to disconnect with the natural world, with what is real.**

The words 'I'm so busy' have almost become a mantra, a disturbing badge of honour, for the lives we lead. Distraction comes in many forms, and work–life balance continues to elude us, as we rush from one thing to the next, packing our 'to do' lists with yet more tasks.

Then there's what we're 'fed' via our computers, televisions and devices – a bombardment of advertising and social media boasts, telling us that purchasing more clothes, a flashier car or a bigger house will make us better, more successful. Happier.

So we rush around, making 'meaningful' connections through our devices, and we place too much value on the things we buy. **Things start to become more important to us than experiences. And that's another cause for our disengagement.**

We get caught up in the daily bustle of living so easily. In my previous business life, I was part of that relentless promotion of 'things', packaging them up to sell to other people. And in my personal life, even if I read about and understood the crisis facing our Earth, I didn't act on that knowledge in any way.

Given this disconnection that engulfs us, it's easy to see how we can think that climate change isn't our problem. But if it's not ours, whose is it? Who else is there to care but us? If we could find that love of our Earth and understand its heartbeat, then we would live differently. I can say this with absolute clarity and absolute honesty because that is exactly what happened to me.

When I walk in a rainforest and witness its splendour, when I feel my connection to Earth, I have pure contentment, uncluttered by the trappings of daily life. All the money in the world can't buy that feeling.

So how do we re-engage with nature?

To be honest, it isn't hard. It isn't rocket science. And there aren't any rules. Just get out there.

Swim in the sea. Go hiking. Let the sun warm your face while you have your morning snack. Go hug a tree. And while you're getting bark in all kinds of places, think about all the amazing ways nature is looking after you. When you start to think about the relationship you have with Earth, remember that it shouldn't be a case of unrequited love.

What are we giving back when we take? How are we looking after our planet when it looks after us? If we love the ocean, should we be using all that plastic? If we love the wilderness, how much are we contributing to landfill? If we love wandering through a rainforest, but we're always choosing our fossil fuel–powered car over public transport … you get the picture. You might start to realise your relationship with Earth is a bit one-sided.

Climate change is the biggest issue we face today. **Changing our behaviour, our lifestyle, and by that I mean profoundly shaking up how we live, is one of the biggest contributions we can make.** To do that, we need to reconnect with the planet and fall madly in love with Earth again.

So let's make a pact to do that.

And while you're at it, slip off your shoes and feel the grass under your feet.

And breathe.

 Pay attention to how much you look at your phone or other screens. **Set limits.**

Next time you take a holiday, **do a digital detox** and leave the phone and screens at home.

Parallel lives

It might sound ironic, but the factors that inspired me to become a cosmetics manufacturer in the first place would later be the same ones that drove me to become a warrior for climate action.

My work as a cosmetics manufacturer stemmed from the belief that **true beauty comes from within.**

With my main brand, I spent a large part of my job talking to women at conferences and lifestyle events, explaining that it didn't matter what you put on your skin – whether it was my products or anyone else's – if you didn't marry that up with inner health, happiness and relaxation.

What astounded me when I had my epiphany and discovered my deep love of Earth was that the same story of holistic beauty applies to our planet.

Looking after ourselves – caring for the inside as much as we care for the outside – is the personal version of how I think we should consider our relationship with Earth.

Everything on our planet is interconnected, including us. And if something along the chain falls apart or is damaged, the whole is affected.

Note: I created four brands. My main brand, for beauty salons, was based on aromatherapy and a holistic approach, which I talk about here; the other three I produced for department stores. I succumbed to the lure of selling to department stores for, what seemed at the time, an enticingly sizable sum of money for creating fast brands using cheap ingredients and heaps of overpackaging. (But that's another story.)

That desire to be healthy and happy, that love of ourselves that we should cultivate from the inside out – this is how we need to feel about our Earth. We can be aware that it needs our help, that it is facing a climate crisis of epic proportions, but awareness is only a surface step.

Think of it this way:

Here's you

Go a week with no sleep, or too many late nights, and you place too much stress on your body. You're going to look terrible, and no magic cream or potion can change that.

Sure, you can fiddle around at the edges and tweak your appearance but, unless you're addressing your body and soul as well (and get some sleep, already), that surface sheen is going to crack.

Here's the planet

We can tend our gardens and put our litter in the bin, but if we're also ripping out trees and polluting our oceans with unnecessary waste, then we're putting too much stress on Earth.

It's like we're giving it too many 'rough nights' – what we're doing to the outside isn't being matched by enough care for the inside.

Get some vitamin nature

Go fall in love this weekend.

It could be the tree outside your house, which provides you with beautiful shade, or the soft grass in the park that yields to your feet.

It could be the unique scents and serenity of a hike, or the glistening salt water of an ocean swim.

Take the time to feel and appreciate it; let it enrich you because, at this moment, you are the luckiest person on the planet.

Find your connection.

Make it personal.

5

Changing how we live

One small step at a time

I've got a thing about tips. You find them in every website, blog and magazine – 10 tips for financial success, 5 top tips for a healthier breakfast – you know the deal. They're neatly packaged, digestible bites of information that offer a seemingly easy path to a goal. But when we're looking at changing the very way we live, we need to move beyond the tips and examine the thinking behind our actions.

<u>Anyone can read a list of snappy suggestions and understand what it's saying, but for that advice to stick, the information has to connect more deeply.</u> Otherwise, we cast our eyes quickly over the tips and then, when life inevitably throws up other stuff to deal with, what we've read soon moves from the forefront of our mind to its recesses.

That's because what we may have registered in our head hasn't travelled to our heart. Once we 'get it' in our hearts, and feel it as part of our own truth, the knowledge becomes a no-brainer, no matter what obstacles come our way.

<u>This is the essence of lasting behaviour change. We make small changes to the things we do, and repeat these new behaviours many times until they become part of who we are.</u>

When we have a strong, deep reason for making those changes – like the love of Earth I've been talking about – then we want to make the changes. In fact, we need to make them, because they reflect who we've become. Once we've made these changes, they're so aligned with our values that they become permanent and profound.

> ✅ **Only 1% of Sweden's household trash ends up in landfill** – the rest is recycled or turned into energy.

It's like when I first stopped buying single-use coffee cups and started bringing my own reusable cup. I used to forget the damn thing all the time. I'd say to myself, 'Well, one more takeaway cup won't make much difference. Next time, I'll remember to bring my reusable one.' That was at the beginning of my journey, when this small lifestyle change was a new addition to my life.

But, as I became more committed, and more used to the new habit, that small change shifted to becoming simply another part of my life.

When that shift happened, there was no more, 'Just this once.' **Now, if I forget my reusable cup, I don't get a takeaway coffee.**

Let's get started

Theories differ on how to change your behaviour. Some people believe that, if you maintain a new behaviour for 21 days, you can create a new habit. I don't know if that's true, but what I do know is that **doing something, no matter how small it is, over and over again gives you confidence. It empowers you.** It shows you, in real time, that yes you *can* do it – because you *are* doing it. Next minute, you're making another, different little tweak to your habits and, almost seamlessly, you're on your way.

For me, reaching that point of true understanding came from taking action myself, and seeing how my actions made an immediate difference.

When I cut down our electricity use, I had no idea it would save me heaps of money, never mind leading me to start a women's movement. I didn't even begin watching my energy use with a view to checking my bill in three months' time. I just looked at our house and lifestyle, asked myself what we were doing that was excessive, and tried to address that.

It all boiled down to stopping and thinking about the way I did things: those tiny habits you barely even think about. Like how I'd walk into a room and, even in the daytime, automatically turn on the light, whether I needed it or not.

Once I noticed what I was doing, I'd open a blind instead to let in as much sunlight as I could.

I hit pause on those unconscious habits. I examined whether they were really serving a purpose – or taking me further away from my energy-saving goals.

From there, I started looking around for other simple things to change, with a new confidence in my ability to shake things up without too much pain.

Where I could, I installed energy-efficient lights, and where I couldn't (for example, being stuck with halogen downlights until energy-efficient LEDs came along), I minimised their use. We used to leave lots of lights on when we'd go out and, with children growing up, we'd always leave a light on in the hall. Now, I make sure that if we leave any light on (so the dogs don't freak out in the dark), it's just a single energy-efficient bulb.

I went through all those **'energy vampires' that were drawing stand-by power** – computers, gaming consoles, gadgets, TVs – and turned them off at the wall when we weren't using them. We had a second fridge sitting under the house that we didn't even need – that was switched off at the wall, too, then was sent off for recycling.

I ditched the dryer (well, it ditched me, but when it broke we never replaced it) and just used good old wind and sunshine to dry our clothes. I discovered other things that made a difference, such as reducing the running time for our pool pump, a big energy sucker.

It sounds pretty simple, but I want you to know that making small changes can really be *that simple*, especially at the start of your journey. So much of this was just about being more careful with the way I used and conserved energy.

> ❗ Never underestimate your power. If 1 million households **reduced electricity consumption by 20%**, it would shut down the equivalent of **two coal-fired power stations**.

Without even knowing it, I was changing my behaviour. I had no idea but, years later, behaviour specialists told me that how I made the shifts in my lifestyle is pretty much the blueprint for lasting behaviour change. And for me, looking back, the piece that made it stick was feeling that deep love of Earth.

There were things I could do straight away that gave me immediate results, such as embracing a reusable coffee cup and always carrying reusable shopping bags. But there were other things that were slower to happen, personal sacrifices that took longer to get my head around and required more planning. It took me two years to really shift my lifestyle away from automatically jumping in the car and towards public transport to get around.

The idea is to be constantly moving forward, constantly considering what you could do next, or better.

Those first easy steps I took to get my electricity bill down were all within my immediate control. What propelled me onwards to each next step was seeing the results. In my case, it was more money in my pocket – I had saved 20 per cent of my household electricity bill.

Ditch the guilt

Most of us know the feeling. It usually comes after watching a documentary about the imminent devastation of Earth, or reading about another species that's been added to the extinction list. You feel a bit sick, but it's not indigestion: it's guilt and fear.

It's a natural reaction – frankly, it would be a worry if it didn't bother you. But it's not a feeling that will keep you going. When you're in this zone, you'll probably try and change everything all at once, then get overwhelmed and throw your hands in the air. I've seen it happen to the most well-meaning people, and it's what I call 'gratuitous behaviour change', instead of genuine behaviour change – it's the hardest to make stick, because it's too easy to let go of it and return to our old habits when a hurdle hits us.

So don't go down the road of feeling guilty; make action and empowerment your new plan of action.

Start small (really small)

So how do we do it? How do we change how we live and make it stick? How do we move from changing which lightbulbs we use to changing our entire life?

The first step is still the lightbulb ...

✅ **The secret of lasting behaviour change (in 10 words or less)**

1. Do something – see a result.

2. Feel empowered by it.

3. Repeat.

In the early stages of changing our behaviours, it's the quick wins and clear results that keep us going. It's not that we need to be rewarded for everything we do; it's only that, if we're just starting out on our journey, those small things that have a quick reward take us to the next thing then the next, because we see the benefit as we go along. Who isn't fired up by seeing something work?

This is where action is so important. As we've seen already, we can't waste our time feeling guilty about what we've done – or haven't done – before now.

What matters is what we're doing at this minute.

Don't worry about what you don't know about climate change. Knowledge, awareness, understanding – all these come along the way. Don't get bogged down in the melting ice caps. There's no time to sit around and feel depressed – that's not going to help anyone.

And don't torment yourself with thoughts of 'What if I fail?'

Focus on doing something, just one thing, and let it move you forward.

Momentum is everything. You must keep moving for these changes to stick.

It's so much easier to start with our own lives, our own homes or our daily routines, and the little things we can do differently in a day.

You know what the other beautiful thing is about these small actions? They show us how much we matter, with every single thing we do. They are uniquely empowering.

I never appreciated this until I got personally invested. And you get braver as you go, as you realise, 'Hey, my world didn't collapse. I'm still having a good time.'

✔ At 1 Million Women, our very reason for being is to shift behaviours and make those new behaviours stick.

We work on the premise that when we show our collective power it helps us maintain our new behaviours.

We demonstrate how, when small actions show immediate results, they move you on to bigger lifestyle changes.

We don't just tell you there's a problem – we offer solutions.

We've learned a lot about changing behaviour

At 1 Million Women, we've worked closely with BehaviourWorks Australia, a behaviour research enterprise within Monash University's Sustainable Development Institute. They conduct research that evaluates and tests new ideas for how best to encourage behaviour change, either for individuals or the social good.

It's a two-way street: the data that 1 Million Women naturally produces, by observing how our community shift their habits, helps researchers understand the challenges of change. Our community of 800,000 women is living data when it comes to this stuff. We capture insights in real time, we collate honest behaviour-change patterns and trends every day, via the hearts and minds of our own community.

Here are some things we've learned about changing behaviour:

1. Taking just one step can have a domino effect. For me (although I didn't realise it at the time), I took up waste reduction as a cause. Once I reduced my electricity waste, I intuitively started to notice what I was wasting in other parts of my life – our food, the way I was consuming. It's really what kept the whole thing ticking along.

2. A little confidence can be enough to keep you going. If you're confident that you can make even one small change and you have belief in your ability to do it, this can be enough to inspire you to keep going. **It's even better if you have a cheer squad to encourage and support you, whether it's family, friends or our 1 Million Women community.** Because it's true: your actions do matter. This is something we focus on at 1 Million Women – because, if you're in the camp of inaction, where you think you can't possibly make a difference anyway, why bother? **It's so important to move from that feeling of helplessness to one of empowerment.**

3. Feeling part of something bigger will help make the changes stick. The power of being part of something bigger, and of seeing the results of your actions as part of the greater whole, will help make your behaviour changes stick. It's one of the most important pieces of our strategy. It goes to the heart and soul of who we are. After all, we're in this together as a community of women.

4. Watch out for the rebound effect – when you make one change but revert to a bad old habit. Sometimes, one behaviour change can be cancelled out by another habit. For example, if I install more efficient lightbulbs in my house, I might revert back to my old habit of not switching appliances off at the wall. Or if I buy a more fuel-efficient car, I might drive more. It's a matter of being really mindful of that trap – **knowledge and awareness are power.**

Turning new behaviours into habits

It's a funny thing but, in those early stages, you'll probably find you love telling people what you're doing. 'Yes, I bring my own cup now … I don't use plastic bags anymore … I've just started a worm farm.' And why not? When we're taking steps to change our lives, we feel good – and we need all the encouragement and pats on the back we can get.

But what you'll notice is that, as these steps become a part of your life, there'll come a time when you won't feel the need to tell people – because it's just what you do. **The real, deep, embedded change has happened when it feels like you haven't actually changed anything. Because it is now normal.**

You can't get there instantly. **You need to see the rewards and the quick wins, and to let them empower you.** I know that's how I built up my confidence and my connection to this new way of life.

Once you're there, your new behaviour will slowly start sinking in as the new norm. But it's a process, which is not always a linear one. Sometimes you fall off the wagon. You just have to get back up and keep repeating that new, better habit over and over, until it feels wrong to do it any other way. You'll know when it's moved from being head-driven to being a part of who you are because it won't feel like an effort, like something you're trying to fit into your life. It will simply be a part of it.

But how do you get to that point? It's a head-to-heart thing.

Head to heart: Internalising your new behaviours

Real, lasting behaviour change isn't about gaining new, scary knowledge and feeling freaked out. It's not about spending hundreds on new stainless-steel straws, reusable cups, produce bags or glass containers (as much as I love all those things). As I explained in Chapter 4, you need to come from a place of deep love for the planet you live on.

Here's how to do just that, step by step:

1. Stop. Notice the way you do things. And be open to how to do them better. Take stock of how you're living. Is that *really* how you want to live?

2. Rediscover nature. Remember how I said you should take off your shoes and feel the grass tickling your toes? Do more of that stuff, more often.

3. Start doing small things that have immediate results.

4. Move on to bigger lifestyle decisions.

5. Get involved in something bigger. Thinking about climate change from a personal perspective is critical, but what keeps us focused is thinking about climate change as our collective responsibility.

6. Think about your legacy. How do you want to be remembered? How do you want future generations to look back on you and your generation?

I'm not trying to depress you with this last point. It's actually a helpful exercise. Studies show that thinking about what kind of legacy we hope to leave behind causes an increase in what researchers call 'helping behaviour', especially regarding environmental causes.

Really thinking about how we will be remembered by people in the future causes us to act in a more environmentally friendly way in the present.

#Leaveitontheshelf

One of our ongoing campaigns is the perfect example of how a quick win can foster long-term behaviour change. We asked people to pledge to make just one simple change when they went to the supermarket: **If you see pointlessly packaged fruit and veg on the shelf, leave it there and choose loose ones instead.**

In itself, that action sends a message to the business that their customers don't want layers of packaging on their food. It's a quick win because it's easy, it's painless and it makes us feel powerful.

Will you take the pledge?

'I pledge to leave pointlessly packaged fruit and veg on the shelf.'

Better still, go to our website and take the pledge there:
1millionwomen.nationbuilder.com/leaveitontheshelf

By leaving these products on the shelf we are sending supermarkets a message they can't ignore.

'Liking' isn't loving – we have to do it all

When thinking about the bigger picture, we do need to keep ourselves in check about one thing. We can donate to environmental organisations, sign petitions, march in the streets (and don't get me wrong, this stuff is important), but we can't do it thinking we love the planet and are doing our bit if, at the same time, our lives are filled with overconsumption. It's easy to think like this, to compartmentalise our lives to the point where it's far too easy to excuse our unsustainable behaviours, because we think we've somehow banked enough karma points by signing every petition that comes our way.

We have to do it all. We have to fight for climate action on every level. We can't go off to events about saving the planet, yet come home and carry on business as usual. We need to make sure our values align, at home and at work. So many of us go off to work at a job that doesn't align with our personal lives. Sadly, too many of us are happy to be loud and proud on social media, just as long as it doesn't affect our personal lifestyles.

Profound behaviour change (the long-lasting habits and lifestyle shifts) isn't easy. But it is possible to rethink how we live, I promise. We're having more conversations about how to simplify, slow down and declutter, to step away from intense consumerism and the consequences we all know stem from that.

Before I really started on my journey, I could talk about climate change at a dinner party and leave all fired up to make a change. But, very soon, I'd slip back into that same old stuck feeling: 'I don't know what to do, so I'll just keep doing nothing.' I was detached from the issue; there was a disconnect from the conversations I was having to the way I was living.

When we're on our journey of changing the way we live, we have moments when we've got clarity and confidence, and moments when we doubt ourselves – it's a roller-coaster. But that's all part of the journey. **Every time you have a win, or you get back up after a fall, you're becoming more empowered and one step closer to looking at life through a completely different lens.**

Think about a long-term relationship. When you first fall in love, responsibility and commitment might feel a little distant – but, soon enough, they make the foundation of your relationship solid. It's the same with Earth. We need to ground our love story with Earth in a firm commitment, taking responsibility for how we live and how we interact with each other and with the planet. **No more chucking food carelessly into the bin because you didn't like the look of it. No more 'just this one' takeaway cup.**

Although changing the way we live is a process with its ups and downs, one thing we can't afford to do is dillydally on our behaviour-change journey. **Climate change isn't waiting for any of us. We need a lifestyle revolution, and we need it quickly.**

Bringing it all together

You want the secret to a happy relationship with Earth? Getting your head, heart and actions to come together. Like an orchestra, where every instrument is hitting the right notes.

The great thing is that action comes in all different shapes and sizes. There are the small things that lead to the quick wins and get you on your way – the reusable water bottle or switching off appliances at the wall. Who doesn't love an easy win like that?

> ✅ The 2017 'Valuing our Clothes: The Cost of UK Fashion' report found that UK households are saving **700,000 tonnes of carbon dioxide a year** by washing clothes at lower temperatures and ironing and tumble-drying them less.

There are bigger decisions that take a little more planning, such as choosing public transport instead of taking the car, or changing your bank to one that doesn't support coalmining.

And then, of course, there are the bigger things that take time to consider, especially where substantial investments are required:

- Will you put solar panels on your roof?
- Purchase a more fuel-efficient car, or an electric vehicle, or get rid of it altogether?
- Will you consciously choose jobs and schools closer to home, to reduce your carbon footprint?

Any one of these decisions, big or small, has an impact and will pull you along to the next step in your journey, as long as you're feeling it in your heart and you're committed to it in your head.

#WomenPower

Three years ago, 1 Million Women launched one of our most successful campaigns, called #WomenPower.

Partnering with the City of Sydney, Australia, we challenged a group of women from different households and lifestyles to lower their electricity by 20 per cent over a three-month period. After all, that number had worked for me and I'd approached the issue as a complete novice! We gave each woman a smart digital device that would monitor her household's energy consumption in real time. During our program, the participants spent the first month getting to know the flow of electricity in their house, the second starting to understand where they could make changes and the third really committing to those changes.

The results were incredible. We were looking for a 20 per cent saving, but the average turned out to be 45 per cent, with one woman managing to reduce hers by a whopping 66 per cent.
I was so impressed, I went on to reduce my household's by another 22 per cent!

We held an event where these women talked about what they'd learned from the experience. The stories they shared, and their achievements, were inspiring and engaging, not least of all because these were women just like you and me. Not climate warriors or energy experts. And all they did was make simple changes in their own households. Here are a few of the things they did:

Nicole (66 per cent), a mother of three teenagers who lives in an inner-city apartment, turned off her under-floor heating, heated towel rails and microwave (plus anything else with a stand-by light) at the wall. She stopped using the clothes dryer and said she'd learned to be happy with the apartment looking like a laundry now and then.

Liz (56 per cent), a mother of two in a large house, replaced her lightbulbs with energy-efficient ones and stopped putting hot food in her fridge so the appliance didn't have to work so hard. She had solar hot water and found that the biggest saver was giving the booster a break and only using water the sun heats up.

Katie (41 per cent), who rents, knew she had no control over electric hot-water heating even after cutting down her hot-water usage, so she focused on wasteful stand-by power, turning off all her devices when not in use.

Lisa (51 per cent) got a foot-pedal power board for her entire media centre, so turning off overnight power was as simple as one click.

Michelle (62 per cent) taught her three housemates how they were wasting electricity, so they could understand and change their habits together.

As well as the huge difference these women made, reducing their carbon emissions by an average of 45 per cent, they saved a lot of money. Overusing electricity by 20 per cent can cost the average Australian householder more than A$400 a year. **The great thing about energy savings is that it costs nothing but will put money back in your pocket** – something we can all do with when electricity prices have nearly doubled in the past five years, and are still rising.

All the women who took part in our program agreed on two things. **That getting others on board, whether their housemates or family, was key to achieving real behaviour change. And that every little action adds up.**

❗ Throw-away fashion is a huge problem.

A survey of 1500 British women found the majority of garments are **worn as few as seven times.**

Distressingly, 33% considered clothes to be **'old' after only three wears** (or fewer).

6

The problem of stuff

The beauty of less is more

I'll never forget the day I suddenly truly understood in my heart that less is more. It was in the early days of 1 Million Women and I was in a meeting, discussing a new campaign we were planning about overconsumption. We were talking about how to get the message out there, and the words 'Less Is More' were thrown around the table.

BANG! That's when it hit me.

The concept travelled from my head to my heart in an instant, and I felt a strange, exhilarating giddiness. (This was happening a lot in those early days of changing how I lived.) *Less Is More.* Wow. I left that meeting with a new-found freedom and an enormous weight lifted off my shoulders. I thought I was already doing a lot, but this was something much deeper. **I was ready to embrace a new way of living that didn't include more stuff.**

> **!** If 1 million women bought their **next item second-hand**, instead of new, we would save **6 million kg of carbon pollution** going into the atmosphere.

High-consumption culture: what we're up against

Of course, changing how we live is never as simple or straightforward as we'd like. Because, when we start looking for ways to change, we quickly realise what we're up against – the high-consumption culture we live in. We're bombarded with advertisements for things that will make our lives better. We're seduced by convenient short cuts, from the plastic takeaways and the single-use throwaways, to the everyday comfort of second and third cars, even bigger houses, the air-conditioning units. We're pushed to buy lots of presents at Christmas, to grab as many bargains as we can in the sales, to mark our status with a new possession, to have the latest fashion, to adopt the latest craze. And if we don't? We're a failure. Or we're tight.

For those of us living in developed countries, we've been told all our lives that buying things makes our lives better. In fact, if you live in Australia, the UK or Europe, or North America, **advertisers spend around US$400 or more per year on you personally, to convince you to buy more stuff. Magazines and commercial television exist to deliver audiences to advertisers.**

Women often bear the brunt of these messages: we're the target market for umpteen products to make our skin hairless; zillions of make-up options for beautifying our faces; a multitude of ways to be a 'better' woman. And they always involve products and services that cost money.

If you're a guy reading this chapter, don't flick past. You're still a major player in all this. Overconsumption is the plague of modern Western culture. No matter your gender, age or sexuality, you're being battered every day with messages to buy more, more, more. And we're buying it, literally: **20 per cent of the world's population consumes 86 per cent of its products.**

When I was doing my researching before I started 1 Million Women I found out that in my own country, Australia, we were buying A$10.5 billion worth of stuff every year that we barely, or never, use. Just let that sink in a minute. Imagine what that figure would be today.

If the entire world lived the way we do in Australia, we would need 4.8 planets to sustain all of humanity, not just the one planet we have.

We can't pass the buck and imagine that someone else will look after it for us. We can't say we want to live more minimally, and go ahead and declutter by throwing 'away' all our superfluous stuff. Because there is no 'away'.

Nothing disappears. It has to go somewhere (probably a landfill).

It can change form – such as the way food waste decomposes and emits polluting methane gas. And maybe you don't see it anymore. But nothing on this planet can just disappear. Instead, we've just put it out of our minds. All we've done is shift the problem.

Instead, we need a new way of living that's based on hope and, most importantly, action.

> **!** The average household has **300,000 items**, and the self-storage market is going through the roof. We have so much stuff that, over the course of our lifetime, we'll likely spend a total of **3680 hours (= 153 days)** searching for misplaced items.

The comparathon

It hasn't always been like this. You probably know an older person – perhaps someone who grew up during the Great Depression of the 1930s, or went through wartime rationing in the 1940s – who only ate mutton once a week as a treat, and remembers owning just one or two pairs of shoes.

And you probably know the story. When our postwar economy boomed in the 1950s, so, too, did consumer products, innovations that were supposed to save us (mostly women) time and energy. And yes, I am grateful I no longer have to scrub my clothes on a washboard to get them clean (thank you, washing machine).

But somewhere, it all went a bit mad. Somehow, we conflated success with stuff – the car I drive, the number of bathrooms in the house I live in, the labels I wear, the bag I carry, the phone I use, the holidays I take. These are the measures of success today.

We're locked into a weird comparathon: that sideward glance we give our friend's new car, that little voice in our head wondering why we're not doing 'as well' as they are. Maybe it's human nature, but it's made so much worse because we've been conditioned to think like that. **From a very early age, there's no space to breathe – everywhere you look, we're being pushed to buy another something, to look better, to *be* better.** In every magazine, on every bus that goes past, even in the sky, we're being sold something.

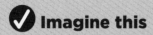 **Imagine this**

Just think of the difference we'd make if entire populations:

- **Bought 50% less stuff**
- **Reduced food waste by 50%**
- **Ate 50% less meat**
- **Drove cars 50% less**
- **Shared 50% more**
- **And if, for 50% of the time, we swapped stuff for experience, or left overpackaged things on the shelf to gather dust.**

Now, imagine if we always used our economic power only for the good of the planet.

What to do when the urge to shop hits

1. **Cuddle an animal.** It's proven to make us happier and healthier (and might just distract you from those cute shoes).

2. **Cuddle a human.** Hugs release happy love hormones that even that sweet 2-for-1 deal can't.

3. **Give something.** Random acts of kindness, like paying for someone else's coffee or leaving a love note for your significant other, make you feel good.

4. **Phone a friend.** Social connection is what makes us happiest, says science (just as long as you're not calling a shopping enabler).

5. **Go read a book.** Your mum was right. You're just bored.

Happiness can't be found in the 2-for-1 bin

We're told the meaning of happiness is having more, better stuff. Who doesn't get joy from possessing nice things? Who doesn't enjoy the thrill of something new?

But what's twisted is this drive for more, more and even more.

Happiness can't be created by *stuff*. But because there's so much gorgeous, shiny stuff on offer in the world – and smart people who have found genius ways to sell it to us – that's exactly what happens. We're never satisfied, because the moment we have a thing, we're being sold the next one, and the first has already lost its lustre. **And quite apart from the mind games, there are the industry games: planned obsolescence, fashion cycles.** It's a cruel joke, and we're the butt of it.

When we're sucked into the cycle of wanting, getting, growing bored and wanting again, we're buying into a culture of overconsumption that's contributing to climate change.

Think about when you've gone on a shopping spree. We're at our 'happiest' when we come home with a whole heap of bags, overflowing with new stuff. We get a thrill when we debut our new purchases. But, maybe, that thrill starts to wear off when we look at our bank balance. When that big credit-card bill comes in, and we don't have any spare cash. When our wardrobe is stuffed to the gills, and we realise we have too much stuff – again. Of course, we want beautiful things, but there's an imbalance happening.

We're buying things as a fix, as a way of getting a temporary high.

It's not easy to not want. And I'm not saying you have to stop buying things. But in our world, there is never an 'enough'. And because we don't see behind the scenes, to the chains of supply and disposal in this consumption, it's easy to think: 'Well, how bad can it be?' Well, pretty bad, as it turns out.

Let's first take a look at fashion – one of the top polluting industries in the world. Then there's cosmetics, where overpackaging is queen. Both industries encapsulate everything that's bad about too much stuff.

The ugly side of pretty stuff

Fashion is delightful; it's an art form. But 'fast fashion' is something else entirely. I'm talking about the stores that churn and burn through trends, spruiking up-to-the-minute clothes for low prices. Some of these are now creating 52 fashion seasons a year.

Spend less, buy more … what's not to like?

I know, it's fun to buy a dress for Friday night on my lunchbreak, one that I'll wear a handful of times before it's a bit too last season and I pass it on to the charity bin. So my new jumper went a bit pilly, but I don't mind because I can get another, better one this weekend. Or, hey, I got a spot of coffee on my T-shirt, but it was only $10, so who cares? Into the charity bin it goes.

Except, only 10 per cent of clothing that goes to charities is accepted. With much of this junk, you can't even give it away!

> **!** In the UK, **7 tonnes of clothing** get thrown out **every 10 minutes**. **In Australia, it's 6 tonnes.**

If that big multinational chain store is producing 400 micro-fashion cycles a month, how can we stay fashionable? If we try to keep up, we'll be churned up in this fashion frenzy, forever buying a new piece because it's 'on trend', only to discard it the next month to follow some new trend.

When it comes to overproduction and consumption, fast fashion is one of the worst. But it's not alone. Cosmetics, cars, homewares, electronics – the model these days is pretty similar, regardless of the product. Built-in obsolescence means a skirt that falls apart after two washes, or a smartphone that's designed to fail just when your contract is up.

Fortunately, there's a bigger push than ever to rethink how we live. The tide is beginning to turn, with **the mantra of simplify, slow down and declutter** offering an alternative to mass consumerism and its consequences.

But how to get started? The answer is easier than you think. There's some more good news, too. **When you do break the cycle and begin to buy less, you're sending a strong message to retailers and manufacturers.**

Breaking the more-is-more cycle

We *can* get off the more-is-more treadmill and break the cycle of overconsumption. It's a matter of taking one small step, seeing the positive result, then taking another. And that first step is not so hard. All you have to do is stop for a moment and ask yourself a few simple questions, the next time you're tempted to buy something new. The answers might just turn you off that shiny new thing. Before you know it, you're buying less stuff, with more money left in your pocket to spend on experiences, adventures or that bill you couldn't pay last month.

1. Will it make me happy?

The first step is to get out of the mindset that a thing can make us happy. Yes, it can give you a buzz, but let's be real. True happiness isn't going to be found at the bottom of a half-price bargain bin.

2. Do I really need it?

Well, do you? Or do you have something similar at home? Has your life been terrible without it? And, even if you're pretty sure you need this particular item, take some time to think it over. Put that throw/mug/scarf/lipstick/new appliance down and leave the store – for a minute, an hour, a day or a week.

3. Where did it come from?

If the answer to the above two questions are Yes and Yes, and you're still thinking about it, ask yourself: where did it come from? We should know the journey of something when we buy it; after all, it's come from somewhere. Is that somewhere a miserable sweatshop, manned by small children denied an education, made with unsustainable materials and shipped halfway around the world? Or is that somewhere a local studio where a small business owner sewed it using vintage fabric? (You can see where I'm going with this.)

4. How long will I keep it?

Okay, has it passed the first three tests? Then here's the clincher. What will happen to it next? Where will it go? What's its next step? Can you lovingly look after it, mend it and pass it along to someone in good nick in 20 years?

If you're still certain that buying this particular item is a good decision, then go for it, and feel good about using your purchasing power for good.

I promise you, we can be just as happy with a quarter of the things we own. We're freer and lighter when we escape from our obsession with stuff.

More is not more, unless we're talking about more love, more experiences, more adventures (and more on that in a minute). More stuff isn't going to make anything better.

Quite the opposite.

Your (anti) shopping list

Before you buy anything new, consider doing these first:

- ☐ **Refuse**
- ☐ **Reduce**
- ☐ **Reuse**
- ☐ **Upcycle**
- ☐ **Borrow**
- ☐ **Share**
- ☐ **Repair**
- ☐ **Swap**
- ☐ **Sew**
- ☐ **Make**
- ☐ **Grow**
- ☐ **Buy second-hand**

The beauty of saying no

Here's another story from my years as a cosmetics manufacturer, which shows just how caught-up retailers and manufacturers are with packaging and overproducing. But it also demonstrates who's the boss, in the end.

Now, anyone who's ventured into the beauty section of a department store knows that packaging is *big*. Dazzlingly swathed products vie for shelf space. Shiny boxed sets, oh-so-cute mini kits and glittering gifts with purchase compete with each other for customer attention, enticing us to buy, buy, buy. As a (now-former) cosmetics manufacturer, I was part of that world.

When I wasn't scrabbling to get products ready for sale, I was busy strategising to pluck someone else's product off the shelves and put mine on. Overpackaging was rife, as was overconsumption, particularly around Christmas time.

Here's what happened behind the scenes. Each December, I'd meet with the department store buyers about which products they wanted from me for the following Christmas. In February, they'd tell me what changes they wanted me to make to the products, based on how successful they had been with consumers in the previous year. I'd always tried to keep my packaging minimal yet beautiful, so every piece was a part of the gift, even its container. **But they'd ask me to put one box in another, wrap it in plastic acetate, then in a plastic tray, all to make it bigger, brighter, flashier.** In June, they'd place orders and, in September, I'd deliver the products.

By the time my products hit the shelves, they'd become less and less recognisable, I'd spent more and more money on packaging, and my margin had been squeezed down to nearly nothing. **Packaging made up 65 per cent of the cost of my products.** I think this must be the case for a lot of cosmetics.

> ❗ Globally, the cosmetics market was worth more than **US$530 billion** in 2017, and is expected to hit US$8 billion by 2023.
>
> That's a hell of a lot of packaging **we have the power to say no to.**

By then, I didn't even care if my product had dropped in quality – all that mattered was that it sold off the shelf. As a small manufacturer, I was utterly beholden to the department stores – they were my lifeline.

But I was not the only one caught in this system. The store buyers are pressured by their bosses to meet a certain target, their bosses are worried about company-wide sales, and so it goes up the line, like a crazy Mexican wave of pressure and stress.

But what I realised from my time in this cycle is that the Mexican wave can work the other way, too. I saw how important the consumer is in all of this. It's the consumer, after all, who is losing out the most, purchasing an inferior product coated in unnecessary packaging that will end up in the bin.

If the consumer left my product on the shelf, and all the other products covered in plastic, that would send a message all the way back to the manufacturer – no to packaging.

So you see? There is an escape from all this overconsumption. It's about our power as consumers. If we leave on the rack or the shelf all that fast fashion, those overpackaged goods or anything else we don't really need, we're sending a powerful message that will change the whole story.

Because, if we don't buy it, they'll stop making it.

Instead, when it comes to fashion (and many other consumer goods), we can create our own style and really own it. We'll be free of the buying frenzy and the pressure because our own style never goes out of fashion.

Because the fact is that less *is* more. **The more value we place on experiences over possessions, the richer our lives will be.** And when we say no to overpackaging and overconsumption, we're marching to the beat of our own drum.

We've got the power.

We just have to do it.

The full beauty regime. You deserve it!

From a young age, we are sold the line that the only way to be beautiful is to buy a whole heap of expensive skincare products.

As someone who's been in the industry, let me share what I think:

We just need to let our skin breathe, use a product or two and marry that up with inner health and well-being.

That's the secret to pure true beauty. I promise you.

7

Redefining happiness

Find joy in experiences, and create memory anchors

If happiness can't be found in the 2-for-1 bargain bin, where can it be found? The short answer is that we need to discover joy in our lives in a way that transcends stuff, so we can look at ourselves in the mirror and say: 'I'm contented.'

We need to find joy and contentment in our hearts and minds, which are never fully satisfied by material stuff, however much the all-pervasive marketing that surrounds us tries to persuade us otherwise.

Nobody wants to live a colourless life, drained of enjoyment. But in the high-consumption society we live in, we've come to expect joy to come in big forms – the bigger holiday, the biggest house – or by accumulating material goods. It's not that we've forgotten the small things – the delicious feeling of sand between our toes, lunch shared with our family, a freshly baked loaf of bread, sunshine on our face. It's just that, for so long, we've been told that we need more and we've come to expect that more will make us feel better.

The reality is that the joy and satisfaction that we've been lulled into expecting by possessing more and more things, never quite eventuates.

Just as overconsumption has crept up on us, so has serial dissatisfaction with having more things than we've ever needed, or even wanted. I feel that previous generations knew these truths, even though, on average, they were far less well-off than we are today. All those wise sayings from my parents' era keep coming back to me. You'll know some of them, too. **Frequent references to 'life's simple pleasures'. 'The best things in life are free.' 'Money can't buy happiness.'**

But if money can't buy happiness, what can? And if stuff can't either, then what?

> ❓ 'Do you get **stressed out by modern-day consumerism** and all of the "stuff" we seem to accumulate in our daily lives?'
>
> That was the question 1MW asked 2000 women.
>
> Not surprisingly, **84% said yes.**

Crazy Christmas

When I was manufacturing cosmetics for the big department stores, at Christmas-time, I would go into the stores that stocked my brands (Kmart, Big W and Target) and stand near my products, to observe who was buying them and how often.

At the time, I was only interested in how much of my product was making its way to the till but, looking back, I was also witnessing the **crazy, mixed-up, last-minute frenzy of Christmas overconsumption.**

So many of us get caught up in this stressful dash to buy gifts.

We hit the shops, throwing stuff into the trolley without much thought about what we're actually buying.

The stress and the following overconsumption hangover – I've been through it myself. And are we getting any kind of long-term joy from the gift giving?

Back then, all that interested me was you buying bucketloads of my product. **Your long-term happiness ended up not being my concern; it was compromised for my short-term gain.**

Joy is a key ingredient in saving the planet

I know it may seem like I've strayed a long way from climate change. But it's all connected. What we consume and how we consume it form a giant loop that leads to the exploitation of the planet's natural resources and the pollution of the atmosphere. Ultimately, we either close that loop, so our individual and collective lives on the planet become more regenerative than destructive, or at some stage – quite likely in the not-too-distant future – we reach a point of human-induced planetary overload. And believe me, there'll be no joy in that for anyone.

If we could just get back to what truly and honestly makes us happy, if we could strip away all the things that we *think* are meant to make us happy, then it would take so much pressure off us. Put the mega challenge of climate change aside for a second.

- **This is minimalism.**

- **This is decluttering.**

- **This is simplifying.**

- **This is downsizing.**

- **And this is saving us money.**

Relationships beat stuff

For the past 75 years, researchers at the Harvard Study of Adult Development have been conducting what is the longest recorded study of human happiness. What have they found?

In a sentence: **'The clearest message that we get is this: good relationships keep us happier and healthier. Period.'**

Not a great job, or owning more things, or making more money, or having the perfect body. Instead, it's our relationships with friends and family that are critical to our happiness. It's important to note that these relationships don't necessarily need to be romantic – they can be friendships or family ties. They just need to be strong, loving and supportive.

Good relationships make us more resilient when life's inevitable ups and downs come along. As much as the advertising industry would like us to believe otherwise, life isn't always meant to be wonderful, or even merely comfortable, and we're not supposed to be happy all the time. Amid the sweetness of life and the high notes, we go through hard times and rough patches, and just plain old boredom. Having more things can't fix that. But great relationships and positive social interactions can put us back on track when we dip too low.

We don't live in a world where everything is great. Instead of wishing it were so, we need to reduce our focus on stuff and relearn what makes us truly contented.

A study of 8500 teenagers conducted by The University of Auckland found that happiness was positively associated with strong connections with **family, friends and school, exercising regularly and eating meals with family.** So, although teens can often wield significant influence over household purchasing decisions, and are subject to the same relentless advertising as adults, based on this study and so many others like it, **their happiness is *not* linked to their consumption.**

Experiences beat stuff

I believe emotional connection is what makes something long-lasting. Too often, with physical things, we outgrow, break or become bored with them.

In my family, we don't buy physical presents anymore. We did in the past, but we've evolved.

Now anything we give each other is an experience – what I like to call a 'memory anchor'. A special meal, theatre or concert tickets, a weekend getaway. These memory anchors are shared by the people who have experienced whatever it is together.

Our memories are tied to these special shared family times. They become part of our lifetime of experiences, our stories.

I feel such joy when I've created a memory anchor with my family, without buying a single material thing. That joy is infectious, because others follow your cues. They get it, once the old paradigm of giving stuff is challenged. You can still have an abundance of joy and laughter and love.

It's just a matter of changing how we think about our happiness, and going back to basics.

Next time you're wondering what to buy someone, consider giving them an experience you can share instead of a product.

✓ Researchers have found that, when it comes to spending, **buying experiences tends to make people happier over time than buying physical objects.**

When test groups made either a product or an experience purchase then were asked immediately afterwards how they felt, their happiness levels were similar.

But down the track, the people who had purchased an experience were even happier about it, while **those who bought products had become less happy.**

Make it mean something

Even happiness can be empty if it lacks substance. To be valuable and lasting, it needs to be *meaningful*.

Maybe we're placing too much importance on happiness per se, when we should be looking for meaning instead – at least, that's what research suggests.

What's the difference? Well, according to the experts, happiness is about taking, and meaning is about giving. **Happiness is about feeling good, while meaning is about contribution and a sense of purpose.**

For 1 Million Women, this is key. Being part of our community makes women happy. This is positive in itself but, more importantly, it adds value to their individual contributions by enabling the power of collective impact.

Achieving even small individual goals can create a sense of achievement and elation – the joy of empowerment.

But there's a multiplier effect when your achievements are part of something bigger, such as a global movement of women uniting to act in their daily lives to combat climate change. I know, for myself, that fighting for climate action has brought more meaning into my life than when I was in the camp of inaction.

Hug a tree – it actually makes you happier

We've talked a lot about how loving Earth is critical in understanding and sustaining long-term lifestyle change. But the great news is that loving Earth also makes us happier, and there are a growing number of studies that show this. Happiness of this kind isn't short-lived, either. Our connection with nature gives us long-term contentment and meaning.

Here are a few ideas to try:

- Bring a plant into your office.
- Grow some vegies. Or just garden.
- Plant a tree.
- Take your children to the park and play in the natural sections, as well as on the play equipment.
- Spend some time sitting under a tree. And if you're so inclined, maybe even give it a cuddle.

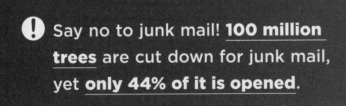

Say no to junk mail! **100 million trees** are cut down for junk mail, yet **only 44% of it is opened**.

What makes 1 Million Women happy?

We asked our members what brought them joy. Here's a quick sample of what they said, in all its glorious diversity:

A **dip in the ocean** on a warm day.

Teaching preschool children about **sustainability** and **waste reduction**.

Losing myself in a crowd at a market.

A meal **shared** with **family or friends**.

Sleeping out **under the stars**.

Waking in the morning to the cacophony of rainforest birds.

A **good long walk**, especially with a friend.

Kayaking and working to **restore waterways**.

Catching a wave.

Thinking of all those **people who are not yet our friends,** simply because we don't know them yet.

Seeing the world move forward on the back of **visionary women who care to take action** makes me **feel happy** beyond measure.

Having a **purpose** that allows me to use my **strengths**.

Listening to music and **dancing spontaneously** to whatever inspires my feet to move.

Picking our **own produce** from the garden.

Cycling, especially **fast downhills.**

Creating something **beautiful** and **plentiful** in my wildlife garden.

More and more people get it, so spread the word

You get it. There's no real limit to the things we can do to gain pleasure and peace of mind, to boost our happiness quotient, so we're no longer dependent on consuming more stuff.

The more committed we are to finding the joy in life, rather than material things, and the more creative we get, the more benefits we'll reap.

The good news is that things are starting to change. More and more people around the world are recognising that material things don't make us happy. For example, the Kingdom of Bhutan established a 'Gross National Happiness Index' to measure the country's well-being as a counterpoint to the traditional, ultra-materialistic economic measure of 'Gross National Product'.

Then there's 'collaborative consumption', the sharing economy where people share, swap, trade, rent or lend goods and services between each other. **'Slow' movements focus on quality, sustainability and fairness for food, fashion and other sectors,** where 'fast' has led to excessive consumption, waste and pollution. There are ethical and sustainable guides and business models for everything from seafood to investments. Micro-lending and micro-payment programs generate socioeconomic improvements in the developing world, often focusing on women as the best conduits for lifting entire populations out of poverty. Not to mention community-based initiatives for clean energy, food gardens, reuse and recycling.

There's an explosion of sharing and second-hand selling, in part thanks to technology, which makes it easier for us to connect. Tool libraries and toy libraries are popping up all over the place. Buy/Swap/Sell Facebook groups are booming.

The list is growing, and we can shout from the rooftops: **'We're making progress.'**

This should make us happy, even though there's still a lot more to be done. We need to share the stories of our own successes and applaud those of others. We need to speak up about the new sources of joy in our lives. Not in a holier-than-thou kind of way, but simply sharing what we've experienced and learned with people around us, who may be leaning towards acting themselves but lack confidence or don't know where to begin.

It's not just okay to evangelise about what we're achieving and the progress we're making in the world around us – it's an absolute necessity. For every single one of us who 'gets it', there are many more people in the world who are still trapped in the cycle of overconsumption.

Find your tribe

I've learned how utterly joyous, rewarding and important it is to find your tribe. There will always be nay-sayers and pessimists, so surround yourself with people who share your optimism, your passion, your curiosity – and your joy.

When you have a tribe, you have people who can shift negativity when you encounter it. They can offer solutions when all you see are problems. They're your people, and they've got your back.

1 Million Women has grown organically as our tribe has united around a common thread. Our tribe is made up of women and girls of every age, across every background and from every corner of the globe, who want to do something within our power to create a healthier Earth for the sake of humanity.

Our 1 Million Women tribe shares each other's challenges and solutions, our insecurities and our ideas for change. We don't all agree on everything. Often there are robust discussions going on, which is how new ideas evolve. But we have a common purpose. We support, encourage, nourish and nurture one another.

Action is propelling our tribe. And women's stories – of their actions and achievements – bind it together.

❓ Take time out with your tribe

Who are the people in your
life who make up your tribe?

Is it time you got together
and talked about the things
that matter to all of you?

Can they support you on your
climate-action journey?

Can you start them on theirs?

8

Be part of the change that's happening

Taking it to the next level

When I launched 1 Million Women in 2009, it was a hopeful time in Australia. Both sides of politics supported climate action, and protecting the environment was among the key election issues that people were concerned about.

Then things began to unravel.

A new conservative political leader, who was a climate-change sceptic, gave precedence to the views of a minority of climate-change deniers over the accepted wisdom of the majority of scientists, and seemed to give them permission to come out of the woodwork.

I'll never forget when 1 Million Women was at Sydney's Royal Easter Show in 2010. We had 120 amazing volunteers working at the stand, and we signed up 3000 women during the show – a fantastic effort. But we also had volunteers running out of the pavilion crying after they'd been verbally attacked. On one occasion, we had to call the police. Up until then, people had been polite and civil, even if they didn't agree with us. But not anymore. Things became tougher.

To be honest, though, these hard times early on taught me a big lesson: **as citizens of this planet, we can't wait for other people – be they politicians or governments – to change the world. We just have to get on with it.**

What change is happening

More and more of us are coming to understand that climate action is about changing the way we live and consume, rather than just being about taxing or trading carbon emissions (or something we leave to the environmentalists to fight). So while we wait for politicians to make their move, people are getting on with it and doing extraordinary things.

We're challenging the overconsumption, waste and environmental disregard that dominated the second half of the 20th century. We don't just think of this as 'greenies trying to save the planet' anymore; we recognise instead that everyone is vulnerable, and that action is about each of us doing our bit.

Live well while protecting the planet

More people are taking on the bigger, more difficult lifestyle changes, finding ways to live well while preserving the planet. It's about saving money, replacing wanton excess with minimalism and simplifying our lives. And, when millions of individuals make changes like these, things begin to shift.

We're having mainstream conversations that simply weren't happening a decade ago, about everything from the clothes we wear to the food we eat and what we put on our skin.

About microfibres, microbeads, unsustainable palm oil, food waste in landfill, the global-scale impacts of eating meat, atmospheric pollution from our cars, plastics in our oceans, and fast fashion growing rubbish piles from yesterday's trends.

Zero-carbon living is now a genuine choice – but when 1 Million Women was starting out, it was a niche idea, easily dismissed as green dreaming. Back then, we were talking about how to have *less* or make the *least* impact on the planet.

Renewable energy is going mainstream

Renewable energy has gone from niche to mainstream. Technology has overtaken policy, filling gaps where governments have failed. **Back in 2009, there were fewer than 15,000 rooftop solar photovoltaic systems in the whole of Australia. By 2018, there were 1.8 million.** That's an extraordinary change, driven by individuals investing in their own solutions.

For the first time, in 2017, global solar generation capacity grew faster than all fossil fuels combined, including coal, oil and gas-fired power stations. The solar and wind industries are creating jobs much faster than the old fossil-fuel energy sectors in traditional industrial economies such as the USA and Australia.

I can't remember anyone talking about home battery storage a decade ago. Now there are tens of thousands of battery systems in Australian households and, like rooftop solar, the uptake of batteries is shaping up to be another extraordinary part of the energy revolution. Going off-grid altogether is something more and more of us can plan for and realistically achieve.

 By 2020, renewables will be the **cheapest form of energy** in the world.

Electric vehicles are also on the rise, which can be powered entirely by renewable energy rather than polluting fossil fuels. A decade ago, only a few hundred electric cars were sold per year globally; in 2017, 2 million electric vehicles were sold worldwide. **It's estimated that, in 2025, almost one in six cars sold on the planet will be electric.**

The world already has most of the technologies and know-how required to achieve this, and innovation is accelerating. But we still lack the consistent political leadership and willpower.

How to be part of the change

By now, I'm sure you've thought of many things you can do in your own life to act. You may have done a lot already. But the key to larger-scale change is how you use your power beyond your own life, to drive the change you want to see around you.

First up, that means getting your family circle involved, from our partners to our kids, from young people to parents. The next ripple out is your community – the local school, your neighbourhood, businesses and groups.

And it works. Great changes in our world often start way out on the margins of society, then gain critical mass as more people embrace them and they make their way into the mainstream.

As much as I'd love for everyone in the world to be on the same page as 1 Million Women, there will always be people who don't want to change. The laggards. The outright deniers. The good news is that only 10 per cent of people need to change how they think and behave for a major shift to take place in the rest of that community.

Change becomes inevitable when the change makers become a significant and vocal minority. It's the tipping point we need, not an outright majority.

> **❗ In 1 minute, how many plastic bottles are bought globally?**
>
> **1 million.**

If we act in our own lives, motivate our communities, use our voices and our votes, we can achieve a major shift and bring about powerful change. Here are just some of the ways we can help kickstart change.

Get the family involved

Do you have a partner who's not so keen to embrace climate action? Or parents who can't quite quit their plastic habit (or maybe don't want to try)? No one wants to be the bore, who lectures people about what they should or shouldn't be doing in a tut-tutting tone. Instead, the most powerful thing you can do is model the behaviours you'd like the rest of your family to pick up. Let them see the changes you're making, and how easy and effective they are. **Talk to them about why you're doing what you're doing, and how much you're enjoying it.** Make it about you, not them – and watch as the cogs start turning over in their minds.

What strategy would hit home for them? If your partner loves a budget, why not outline the economic benefits of energy savings, or of switching banks to get a similar deal with more environmental care as an added extra. If your loved ones are partial to a takeaway coffee, why not buy them a beautiful reusable cup and tell them about a local café that offers discounts for using it?

Our surveys at 1 Million Women repeatedly show that one of the big reasons behaviour change doesn't stick in a household is that the rest of the family isn't on board.

So why not take on a climate-action challenge as a family and make it fun. It might be Plastic-free July, or a zero-waste week. Maybe it's getting your electricity bill down by 20 per cent, then using the money you save to do something fun together that they'll enjoy and remember.

Make it all about small wins, inclusivity, and heaps and heaps of fun. You can't go wrong!

Get young people involved

Early on at 1 Million Women, **we made a commitment that young women, under 18 years old, should make up 20 per cent of our community.** We're well on our way.

Young women belong to the highest consuming generation in the history of humanity, simply because everything is at their fingertips. But, on the flip side, it is their generation which is the influencer and leader of not only today, but also the future.

Teenagers are passionate, and their voices need to be included in the conversations we're having. At the same time, we need to make them more aware of how the world works and the way households and communities operate.

We need to prepare today's generations of young people and our children to become extraordinary stewards of our planet.

We need to give young people power and agency.

Talking to young women – and being scooped up in their enthusiasm, hope and courage – is one of my favourite things to do.

Get the kids involved

I hope the next generation of kids will look after our Earth because it's just part of who they are and how they live. It's up to us – as parents, grandparents, caregivers, relatives and friends – to make that a reality.

If you have kids in your life, make being eco-friendly the new normal so they grow up knowing nothing different.

Here are a few ideas:

- **Help children to grow something, for example, herbs, flowers, vegies, even a tree.** Show them what Earth gives to people and talk to them about how, in turn, we need to give back by looking after it.

- Go on a bushwalk together. **Encourage them to listen to the birds, touch the trees, breathe in the air and feel the water.** Teach them how to love the planet from a young age.

- **Play games to teach them about avoiding waste.** Can we give this thing a second life? If we need to throw it out, which bin does it go in?

- **Give them responsibility for the water they use.** If they love using the tap (many kids do), encourage them to turn it off and 'save water for the fishies'. Ask them: are you a water waster or a water saver?

- Same goes with lights: **make it their 'job' to switch off the light when they leave a room**, or to make sure all the power points have been turned off at the wall when you leave for the day.

- Instead of taking them to department stores to buy new clothes, **browse in op shops or flea markets** to give pre-loved items a second life.

- **Play dress-ups – for keeps.** Host fun clothes swap parties for your kids and their friends.

- **Don't overdo it on the 'stuff'.** Is there a better option, such as second-hand, homemade or experiences, rather than more stuff? We should be living to exactly the same principle as adults.

- **Take them to the library to learn about borrowing instead of buying**, and find some fun picture books with an Earth-friendly message while you're at it.

- Keep talking to them. **Kids are smart: they notice what you do and what you say.** Get them involved.

Climate Cooler Kids

Before I started 1 Million Women, I wanted to get some early runs on the board. I wrote a little book called *Climate Cooler Kids*, which followed the story of Edwina, 'the coolest kid on the planet', as she showed her friend Charlie easy ways he could help the environment. With a small team of volunteers, we'd visit kindergartens and preschools, read the book to the children and engage them in activities inspired by the book. Then we would ask them to do a few key things when they got home:

- **Turn off the tap when they brushed their teeth.**
- **Feel the shade of the tree (to encourage a connection to Earth).**
- **Turn off the light when they left an empty room.**
- **Tell their parents they want them to join in, too.**
- **Put buckets in the shower, and use them to water your garden.**
- **Check if the dishwasher is full before turning it on.**
- **Help Mum and Dad to hang out the washing, and see how the sun dries it like magic.**

If you did those things, you became a *Climate Cooler Kid*, and you got a little tree to plant. I read the book to hundreds of preschool kids, aged four to six. They loved it!

When I'd pause in the book to ask the kids a question, they'd shout so loud it hurt my ears: 'Mr Brown's a water waster!'

They'd tell me how much they loved Earth and how they were willing to do whatever they could to look after it. I think the roles reversed – these kids became the teachers. Their boundless enthusiasm helped us, the educators (me in particular), who were uncertain about how to approach climate change, to move from the camp of inaction to action.

Are you *that* young woman?

If you're that young woman reading this, then I know you're going to keep showing up in this fight. There's a new way I would love for you to get involved, beyond changing your own lifestyle (although that, seriously, is so important).

I'd love for you to become a 1 Million Women Ambassador and take our message – and our new app – to your school.

We'll help you put together an assembly presentation, give you all the tools you need and show you how you can turn your friends (and teachers) into everyday climate activists.

Just jump on our website and follow the prompts to becoming an ambassador to find out more.

Motivate your community

It's amazing how quickly you can get more people on board, just by talking to them about the issues and the better options that are available to all of us. If you're at work, talk to your colleagues and management about whether you could set up a waste-watch committee to look at ways to make your workplace more planet friendly. **As well as reducing waste, you could save your organisation money.**

Here are just a few ideas:

- If you're not doing it already, introduce recycling.
- Ditch the coffee-pod machines. Coffee pods are a single-use nightmare!
- Encourage everyone to turn off power-hungry computers, printers, microwaves and other devices when not in use.
- Encourage colleagues to walk, cycle, bus or train it to work rather than drive. Or carpool (if you really must use the car). You could even introduce a walk-to-work week.

If you're part of a school community, talk to parents, carers and administrators about ways to reduce waste in the school, and motivate and inspire the kids to take climate action.

Sports clubs, local shopping centres and community centres are other communities just waiting to be inspired into climate action by someone like you.

Raise your voice, and cast your vote

It's disappointing, unforgivable and immoral when our politicians are so slow to act. Some are better than others but, overall, they aren't acting at the speed that's needed. The truth is, though, if enough of us demand change (just 10 per cent, remember?), we can shift the system. If enough of us insist that there be no more coal-fired power stations – and we won't vote for a political party that doesn't support renewable energy – the politicians will listen and will act.

Governments are motivated by the pressure we put on them. I've witnessed it. I've learned it.

If enough of us put pressure on our politicians, from local levels to the national governments, change will move faster.

Remember: you're important. Individuals make up households, communities, small and big businesses, schools and electorates – every single part of society starts with an individual. **We need to do everything we can while asking our government to do more, too.**

Ignite your inner activist: lobby and apply pressure to your local politicians. Make your voice – and your vote – heard, and remember, your voice becomes so much more powerful when you are taking action yourself.

And don't stop with the politicians – use your power to make waves. Here are a couple of target areas to create major impact:

- **Banking**: Swap to a bank that doesn't invest in fossil fuels, and tell your old bank why you left them.
- **Shopping:** Support products that are doing the right thing, and leave the other stuff on the shelf to send a direct message to the brands and manufacturers. They will quickly get the point that we don't want them.

Over to you

Now it's time to ask yourself, honestly, are you doing all that you can? We need to be honest with ourselves. I know it's a journey, and I know these are seismic shifts that I've been talking about. But we need to challenge ourselves: **What more can I do right now?** Then next week ask again: **Am I doing all I can? Or can I do a bit more?**

It doesn't matter how you get involved, or how you start your journey. But once you're in, don't let this be a passing fad or casual pursuit. Really feel that love for Earth. Embrace less is more. Make those big lifestyle shifts. Push governments and businesses to do the same. Use your voice. Make your vote count.

Because climate change is an urgent issue, and we can't be fluffy about this.

You are powerful.
You are a change maker.
You are the most powerful part of this story.

So start small or start big and, if you're already well on your way or you're a climate warrior from way back, remember we're all in this together and we all need to keep going. If you're at the beginning of this journey, I promise you that, if you're in this wholeheartedly, you can't stop at one thing. One change will move you along to the next one. You'll find your voice, you'll influence others and you'll do it all. Trust me. I've lived it. I see it happen for others every single day. This is not just about getting your electricity usage down, or your food waste under control. It's about exercising every aspect of who you are in this fight and this story. Acting will empower you.

I've always believed in the power of a single voice, a single vote, a single choice. One voice multiplied by millions is an almighty shout that's hard to ignore. That's been my vision for 1 Million Women from day one.

Your money, your voice, your vote, your everyday habits –
every choice you make can be a choice for climate action and
climate hope.

6 steps to political change for climate action

1. Who are your local politicians? Find out.

2. Do they support stopping all new coalmines from being built and getting to 100 per cent renewable energy as fast as possible?

3. If they do, give them all your support.

4. If they don't, tell them they won't get your vote unless they shift their support to REAL climate action. Request a meeting. Send a letter.

5. OR change your vote. Use your vote to endorse someone who has climate action at the top of their agenda.

6. Be noisy. Have conversations. Tell your friends, family, your networks, your work colleagues and social media acquaintances to do the same.

For our planet, for the people we love and for the sake of all humanity, we must elevate climate action as a top priority come election time.

PART 2

SAVING
THE
PLANET

Toolkits

9

Energy
toolkit

We can all
be part of a
clean-energy
future

I have a passion for saving energy. For me, this is where my behaviour-change journey began. My epiphany and founding 1 Million Women all started with my household electricity. Who would have thought that this seemingly easy action – getting my household's electricity use down by 20 per cent – would have had such an enormous impact? When I got our electricity bill and saw how much money and pollution I'd saved from the small actions I took, I knew I was powerful.

Energy is simultaneously one of the world's most daunting climate-action challenges and one of its greatest opportunities. We can be optimistic that real change is happening already, and there's more to come.

The ultimate vision is a 100 per cent renewable-energy future reliant solely on wind, solar, hydro, geothermal, tidal and other clean-energy technologies – including optimal energy efficiency, because the cleanest and lowest-cost option is the electricity we never use.

Globally, households are responsible for nearly one-fifth of all energy-related greenhouse gas pollution, so we all have a part to play in achieving a clean-energy future.

Your part in the extraordinary energy transition now underway is vital, achievable and can reward you with lower electricity bills and a lower carbon footprint. Immediately. The starting point – being more efficient with energy – you can do without spending a cent and still receive dollar savings and pollution reduction.

Saving 20 per cent off your electricity bill is realistic for most households in countries such as Australia, the UK and the USA. (That's how I started.)

It's something that you and I, and everyone in the house, can do easily. (The key is to have everyone you live with on board – that gets the best results.)

Remember: every home is different. So much of this has to do with your household's habits and understanding the flow and rhythm of your home. **So much of energy savings is about being more aware, more careful with the way we use and conserve energy.**

In terms of your whole journey, the first step is an easy one. And when you're starting to change how you live, you want it to be easy, don't you?

Energy efficiency is a big one

In Australia, households are responsible for at least one-fifth of the country's total greenhouse gas emissions, that's more than 18 tonnes per household each year. On average, almost 40 per cent of this – approximately 7 tonnes – comes from the electricity they use.

With electricity and gas prices surging in recent years, energy expenses are a significant and recurring component of total living costs. In the past 60 years, houses in Australia have nearly doubled in size – on average they are among the largest in the world – which locks in higher energy consumption, and the number of electrical appliances and devices in our homes has increased dramatically.

There is so much we can do to change this story, from picking the low-hanging fruit actions that cost us little or nothing, through to bigger lifestyle and investment decisions, such as upgrading to highly energy-efficient lighting, adding insulation to our walls and ceilings, double-glazing windows, purchasing 5 to 7 energy star–rated appliances and installing rooftop solar panels and battery storage systems.

Saving energy does a lot more than save money. When most of our electricity still comes from coal-fired generation (which is still the case in Australia, at about 75 per cent), we must remember that it's our responsibility to limit our energy usage as a way to help the planet.

The NO CARE, NO RESPONSIBILITY era is over.

In the past, many households didn't pay much attention to wasting electricity, because:

1. Energy prices were low (I remember I used to fear the phone bill and laugh off the power bill. Well, now that's been reversed).

2. We had fewer electrical appliances and relied less on electricity.

3. We were less aware of how much greenhouse gas pollution was being emitted from coal-fired power stations, and also how much environmental damage it was causing.

Get to know the energy flows in your home, then start by focusing on the things you can change straight away without spending a lot.

❓ The average Australian home has about 60–70 electrical appliances, gadgets and equipment – and a lot more if you include every lightbulb. These are all 'energy actors'.

How many appliances do you have?

How many can you turn off when not in use?

How many can you do without?

Here's a plan to get you on your way.

Switch off appliances at the wall when not in use

Microwaves, phone chargers, computers, lamps, TVs, stereos, game consoles – turning off electronic equipment is a simple way to make a big difference.

Around 10–15 per cent of household power use is drawn from appliances you think are switched off but are actually on stand-by, continuing to use power. If it has a light on, it's in stand-by mode.

Make sure you turn off everything at the wall.

Install energy-efficient lights

Solely the use of the lights in your home can cause more than 1 tonne of carbon pollution each year.

By switching to energ-efficient lights such as LEDs, you will reduce pollution from lights by 80 per cent, compared to old-style incandescent bulbs. And halogen downlights are real energy-sucking villains.

If your budget is tight, perhaps start by replacing one to four incandescents or halogens with swap-in LEDs each month until your whole home is done.

Buy energy-efficient appliances, buy second-hand or don't buy at all

Does the appliance you're looking to buy have a high energy-star rating? It's often worth paying a little bit more upfront for the purchase and then earning some of that back by lower operating costs. Always look at the ratings: 5 stars is good, 7 stars is excellent – the more the better. Also, study the appliance manual so you know which are the most energy-efficient settings, and use those as your default operating modes.

Consider renting or buying second-hand appliances, as this will save resources, contribute to the sharing economy and save you money. But be wary of old, outdated, high-energy consumption clunkers.

Ditch the dryer

A clothes dryer accounts for approximately 12 per cent of electricity use in a typical household. This is massive. By hanging your clothes on the line or a drying rack, rather than using a dryer, you can save more than half a tonne of carbon dioxide pollution a year. And drying with the sun and a breeze is 100 per cent FREE.

Try not to have a wash day when it's raining and, when you do wash, use cold water for another substantial saving.

 If 1 million of us kept our showers to **less than 4 minutes**, it would have the same impact as **planting 900,000 trees.**

Take shorter showers

Place a timer or a little note in your shower to remind yourself to get out after 3–4 minutes, and make sure to install a low-flow showerhead.

In addition to saving water, having a shorter shower reduces the energy required to heat the water.

Don't use your air conditioner or heater

About three-quarters of Australian homes have at least one air conditioner. **So much energy is needed to run these appliances that summer heatwaves can cause peak-demand overloads for the electricity system.** In countries with colder climates, heating appliances are the norm rather than cooling, but the energy demand is similar.

There are many ways you can keep your house cool in summer and warm in winter without having to switch on your air conditioner or heater:
- Keep your house cool by learning how to properly ventilate.
- Block out the sun.
- Use fans, which use much less energy than air conditioning. But, remember, a fan is only effective when you're nearby to feel the moving air, so turn it off when no one's in the room.

In cold weather:
- Keep yourself warmer by layering up (dress for winter).
- Retain warmth in your home by closing windows and doors.
- Eliminating cold draughts with door snakes and seal gaps in windows.

When it is really needed:
- Set the thermostat on your air conditioner one or two degrees warmer and your heater one or two degrees cooler. **Every degree can save you 10 per cent on running costs.**

Improve your fridge's energy efficiency

Here are some things you can do to ensure your fridge is running efficiently and to help it retain its stored 'cool':

- Fix damaged seals.
- Close the door promptly.
- Leave space around the fridge for proper air circulation.
- Set the fridge temperature to 3–4°C.
- Clean air filters regularly.
- Don't overstock or understock it.

Get professional help or use monitoring tools

If you're prepared to invest modest dollars upfront to benefit from ongoing savings over time, have a professional energy audit done by an expert and implement their key recommendations. Or install a real-time energy-monitoring device to see power consumption as you use it.

Data will help you determine what's using the most electricity, and where you may be able to realise savings by making behavioural changes and switching to energy-efficient equipment.

Actually use it

Through inattention, it's easy for an estimated 10–20 per cent of the energy we pay for to simply be wasted, which adds to the pollution load and costs us money for no useful results. Rooms don't need to be lit, cooled or heated when there's no one in them.

Anything that heats or cools is likely to be doing more 'work' and using more energy than merely providing light or sound. So, pay more attention to these appliances, especially anything that runs hot (like those old halogen downlights that waste energy while producing excess heat, which also adds to cooling costs).

Engage your household

The number of people living in your household is the single biggest variable factor influencing your electricity consumption, and the habits of every individual affect your bill.

If you really want to save power and money, you need everyone on board.

One innovative approach we've heard was for a family with older children living at home:

- The parents said their kids didn't have to pay any rent as long as they paid the electricity bill – and energy use went down!

Energy stars, eco settings and e-waste – oh my!

Appliances can account for about 30 per cent of a home's energy use. Here's a short guide to help you understand energy ratings and the features of certain products that make them energy efficient.

What to consider before purchasing an appliance

Does the appliance have a high energy-star rating?

The energy-star rating gives a comparative assessment of the model's energy efficiency, and provides an estimate of the appliance's annual energy consumption. The more stars the better.

Does the appliance have eco-friendly functions?

Read through the instruction booklet online to ensure you're aware of the different functions and options available for your potential purchase, and check if the product has energy-efficient functions. For example, some dishwashers have a half-load option. This feature uses less water and electricity, which will save you money.

Buy the product that suits YOUR needs

While we are often tempted to buy the biggest and fanciest appliance – particularly when it's on sale – ask yourself, do I really need it? Bigger appliances could drive up water and electricity bills. Instead, choose a product that's sized to your typical needs. Importantly, also consider going without!

What about the waste?

Also consider the impact of your purchase on the waste stream. Instead of buying a new product, you could purchase second-hand and extend its life, rather than it being sent to landfill. Make sure to recycle your old appliances safely. For example, electrical waste (e-waste) contains hazardous materials and should be disposed of in specialised recycling programs rather than going to landfill. Many countries have product stewardship schemes that make manufacturers responsible for ensuring e-waste is recycled, and rubbish tips often have designated e-waste collection points.

Power and electronics

Electronic gadgetry is a major factor in home-based work and study, as well as entertainment and leisure activities in modern lifestyles. Every device uses electricity and is an opportunity to save on energy.

1. **Switch off TVs when you're not watching**, which will do more to reduce energy use than almost any other home entertainment device.

2. **Make sure the brightness level of your TV is right** for your room, as the factory settings are typically brighter than necessary. Also, make sure you switch on the ambient light sensor – if you're watching your TV in a darker room with the sensor switched on, it can dramatically reduce power consumption by adjusting the contrast of the picture automatically.

3. **If you're buying a new TV:**
- Think about the size and type of screen you choose. An energy-efficient 80-cm LCD screen model will typically use half the power of a 110-cm plasma screen. In general, the smaller your TV, the less it will cost you to run.
- Research energy star information to be sure you are buying a TV with optimised energy-saving features and built-in energy efficiency.

4. **Switch your TV to energy-saving mode**, rather than using the normal viewing setting. This usually dims the backlight, which means power consumption should drop by about a third.

5. **A laptop computer is more energy efficient than a desktop and monitor set-up.** Laptops, desktops and monitors are all becoming increasingly energy efficient.

6. **Computers effectively use similar power whether they are busy or idle.** If you leave them doing nothing, they are using almost as much power as if they are number crunching or accessing information, which is why 'sleep' mode is so useful.

7. **Don't forget to switch off your computer and any peripheral devices, such as your printer and scanner, overnight.** Using a power board to plug in several pieces of equipment makes it easier to switch them all off with one action.

✅ If every household in the USA **unplugged electrical devices** when not in use, it would **save 44 million tonnes of carbon pollution** from entering our atmosphere.

Power and light

Having the 'service' of light when we need it is one of the greatest benefits of electricity in our daily lives. And every light is an opportunity to save energy.

1. **Only turn lights on if you need them.** Daylight is the most energy-efficient form of lighting.

2. **Turn off lights when you leave a room**, unless you are going back within a few minutes.

3. **Choose LED lighting options** wherever possible because, even if the upfront cost is higher, they'll save you money over time and cut pollution.

4. **Avoid installing halogen lights**, especially the low-voltage downlight types, which go in recessed fittings in ceilings and walls. Often these can be swapped out for LED alternatives.

5. **Think of lightbulbs as one-off purchases** that last many years, like the lamp itself, rather than something you simply throw away after a short time.

Weatherproof your home

Heating and cooling systems are some of the major sources of emissions in most homes. Using them wastefully not only burns energy, but also money. **Taking simple steps such as draught-proofing your home can reduce your energy use by 25–50 per cent**, which is good for both the environment and your wallet.

Draught-proofing

Draught-proofing is an energy-efficient way to help keep the heat out in summer and the warmth in during winter. However, there is only so much a door snake can do to maintain a comfortable temperature indoors. Many cracks and gaps often get missed or ignored, because they're just too awkward to fill. To seal gaps around windows and doors – major energy-wasting culprits – try using expanding foam, which cures to a hard consistency and can be cut, sanded and painted for a clean finish, or other sealing products from the hardware store.

7 ways to stay cool in the heat

1. **Shut the windows and pull down the blinds in rooms you won't be using, before it starts to get hotter outside.** This will block out the sun and help to keep your house cool throughout the day.

2. **Open your windows and blinds for ventilation once it gets cooler during the evening, and only if the temperature outside is cooler than inside.** Do the reverse if the objective is to warm things up rather than keep them cool.

3. In the first instance, **use fans in your home to keep yourself cool.** They use much less energy than air conditioning – in fact, the top driver of increased electricity demand during heatwaves is air conditioning.

4. **Take a look at the thermostat.** If you use air conditioning, it's best to target a temperature between 23°C and up to 26°C, rather than between 18°C and 20°C. Keeping the thermostat at these levels could save 15 per cent or more of the air-conditioning running costs. It's also wise to zone your cooling to where you spend time, rather than chilling your whole home.

5. **Turn off electronics!** Computers, TVs and other electronics generate heat when sitting idle or even when on stand-by, so switch off devices at the wall when they're not in use.

6. **Don't run appliances such as dishwashers or washing machines during the day** – unless you have solar panels and aim to maximise solar use in the home, which will save you money and cut pollution. For non-solar homes, save these activities for overnight, after evening peak periods, and only when you have a full load. Running appliances in the daytime increases the humidity in your home, which heats things up and will make your air conditioner work harder if it is on.

7. **Consider installing shading, ventilation and insulation** for the long-term, or plant leafy trees to provide shade. You can retrofit existing homes with ceiling insulation, draught stoppers and shades to stop sunlight from striking windows directly, stopping the heat from entering in the first place.

7 low-cost ideas to keep warm in winter

1. Heavy curtains are a really effective way to insulate windows. As many open-plan homes don't have barriers between certain rooms, which makes it difficult to keep them warm, simply tack a blanket or curtain to the top of a doorframe or hallway entrance to solve the problem temporarily.

2. Put an extra blanket under your bottom sheet to form an insulating barrier between you and your mattress. Even if you already use a mattress topper, the extra blanket will add another level of warmth.

3. Keep doors to unused rooms closed. Heat will escape if you leave too many doors open. Try to have everyone in the house use only a few rooms that are close to one another – say the lounge room, kitchen and dining room. That way, you can all stay together, and your body heat will stay in the rooms you're in.

4. Close your windows. A window that is open, even just a tiny gap, can cause your entire house to freeze. Make sure the windows stay sealed. If you're worried about your house getting stuffy, open a window for a short period during the day so your house can be aired out.

5. Rugs, rugs, rugs! Who doesn't love a nice rug? If your house is tiled or has floorboards, a rug helps retain warmth – and it will keep your toes warmer, too, if your slippers have gone astray.

6. Stop cold draughts from entering through doors and windows. Use a good old-fashioned door snake or, for a quick fix, try running a towel along the bottom of a door for a makeshift draught stopper.

7. Keep yourself warm. Layer up and dress for the cold, or reach for a hot-water bottle.

The renewable energy story

Just a decade ago, the idea of having solar panels on your roof, generating clean electricity from the sun, was expensive and niche, far from the mainstream. That story has changed. Now, it's something people around the world can do – and are doing.

Over 1 million US homes, 1.5 million in the UK and the world-leading 1.8 million Australian households have joined the rooftop solar revolution.

Solar is now a major global industry, and there is plenty for you to research. If you'd like to go solar, there are many commercial and other sources of information available.

If you already have solar panels, you may want to get more or add battery systems for energy storage.

These are significant financial decisions, so get expert advice on what's right for your home.

The more of us taking action ourselves to move to renewable energy, the sooner we'll get to a zero-carbon pollution future.

✔ A beautiful thing about **off-grid solar** is that, almost automatically, you find yourself **more in tune with the world** around you.

You're aware that a sunny summer's day means lots of power (and time to get the washing done) and that in winter, and when it's cloudy or rainy, you need to be more frugal.

How to speed up the renewables movement

☐ **Make sure your bank and superannuation or pension fund support investment into renewable energy.** If they don't, move to ones that do (see Chapter 14: Money toolkit).

☐ **Find out what your local politicians are doing.** If you don't think they're doing enough, let them know – and inform as many people in your community as you can, too. Pressuring your politician to do better and to put support for renewables on the top of their policy priority list is critical.

☐ **Use your vote and your voice.** Show politicians that you won't vote for them if climate action and a clear plan to fast-track renewables aren't on the top of their agenda.

☐ Whether you own your home or are renting, in Australia and many other countries **you have the power to switch your energy company**, so you're giving your money to one that (for the most part) provides renewable energy, or is at least carbon offsetting or doing a combination that advances renewables. Watch out, though, when doing your customer research. **Be wary of energy companies that are merely 'greenwashing'**, using marketing spin to falsely claim their practices help the environment. Look for expert advice such as the Greenpeace Green Electricity Guide.

☐ If you're a home owner, **consider making the switch to renewables.** For an investment of a few thousand dollars, you can install solar panels to harness the power of sunlight and reduce your reliance on the electricity grid. Or, one day, even go off-grid altogether.

☐ **Don't feel like you can't do this if you live in the middle of a city.** Australian sustainability guru Michael Mobbs (aka the 'off-grid guy') is doing this right now in the middle of inner-city Sydney. Meanwhile, there are many off-grid communities popping up all over the world, and they often welcome visitors who'd like to spend time there and help with upkeep, while at the same time learning how to replicate the same model. Whether you want to go all-out or simply start with a few solar panels, it's all progress towards a clean energy future.

☐ If you already have solar, it's important to **keep track of how well your array is performing**, and maximise the value of your investment by using as much solar as possible in your home. Otherwise, you might be surprised with high electricity bills if your panels aren't working as expected, or you're importing too much from the main grid. **Use a monitoring system, know the wattage of your appliances and the best time to run them** – for example if you have a pool, run the pump while the sun is out. **Make sure your panels don't have shade covering them and they're kept clean.**

☐ With residential storage batteries for solar power now more available and coming down in price, **storing the energy you generate is becoming more practical.** This, in turn, may allow more people to go partially or completely off-grid.

☐ Importantly, **cutting yourself off from the grid entirely needs careful planning**, so get good advice and plan each step.

10

Food toolkit

The journey of food

Rewind to 10 years ago. I liked having a full fridge and packed pantry (despite our small kitchen). I'd fill up a big bowl on the kitchen table with tomatoes (more than we could eat because they were also for show – the colours picked up those of the big painting in the lounge room). We were a chaotic family (those of you who know me will be nodding your head): no proper dinnertime, no meal planning.

It wasn't unusual at our house to have 20 around the table. There were big lunches that spilled into the night and big brekky cook-ups for whoever stayed over. The table would burst with food, without much thought on portion sizes. I didn't *want* to be wasteful – I just never thought about how much food was actually going into the bin (a lot) or where it went after that.

Sounds crazy?

The fact is that **Australians waste at least 20 per cent** of the food we buy – that's $1 out of every $5 spent on food shopping.

Once I started looking at ways to reduce our household food waste, I got a big fat wake-up call. Turns out those 'harmless' food scraps I chucked in the bin on a regular basis were making their way to landfill, where they broke down and released methane. **As a greenhouse gas polluting the atmosphere, methane is 23 times more potent than carbon dioxide.**

! Imagine **paying for 5 shopping bags** filled with food at the store, then walking outside and immediately **throwing 1 bag in a rubbish bin**.

Even worse, all the resources and emissions (the embodied energy) that had gone into producing that food and delivering it to me were being wasted, too.

So, honestly, cutting food waste in our family was kind of easy. There was a lot to work with. I began to think about food as a journey: from when and where I go to the shops, to what I can grow myself, to how I'll manage leftovers. I bought a couple of worm farms. I planted the herbs that we always cook with. I began to plan meals before I went shopping and, instead of just filling my trolley with random food, I'd ask myself: *Will it all be eaten? How much waste will there be?*

I love the joy that comes from cooking, the freshness, the creativity, the conversation and the laughter that comes with sharing food. It truly nourishes my soul. The good news is that none of that has to change.

Why are we wasting food?

- We buy too much.
- We cook too much.
- We don't have a plan for leftovers.

Buy what you need. Eat what you buy.

We can avoid wasting a massive 60 per cent of the food we throw out simply by managing it better.

When we waste food, we waste not just the food itself, but also the time and resources that went into producing, delivering, selling and preparing that food.

- Don't be seduced by 2-for-1 deals when you won't use that much before it spoils.
- Don't go shopping when you're hungry – you'll always buy more than you need.
- Stick to a meal plan.

> **❗ More than 1 million edible bananas are thrown away in Britain every single day.**
>
> Old and bruised bananas can be blended into a smoothie, baked into a cake or frozen into ice-cream.

Tackling food waste

Reducing food waste is a constant focus for us at 1 Million Women. Recently, we conducted a survey to learn more about our food-waste habits; we had 13,000 respondents.

One of the questions we asked was to identify the main food that gets wasted in their homes. I doubt you'll be surprised to hear that the answer was fresh fruit and vegies.

When asked what their main reason was for throwing out food:

- **54%** said it gets left for too long in the fridge or freezer.
- **25%** said family members don't finish their meals.
- **20%** said food goes off before the use-by or best-before dates.

When asked what food-waste measures they had in place:

- **80%** said they use up their leftovers.
- **78%** said they check the fridge before they went shopping.
- **77%** said they shop to a list.
- **60%** said they buy only what they need.
- **59%** said they consider how the food is stored to keep it fresh.

When asked what else they do to manage food waste:

- **53%** said they have a compost heap.
- **31%** said they feed it to their pets.
- **18%** said they had a worm farm.
- **13%** said they had chickens.

Now, how about you? Consider how much food you throw away:

- An excessive amount?
- A reasonable quantity?
- Very little?
- None?

Why food waste is important

The food and beverages we consume can account for a third or more of our total environmental footprint. Australian households, on average, waste food worth approximately A$1000 every year, which also means wasting everything that went into bringing that food to you – energy, water, nutrients, carbon emissions and more. For example, Australians waste more than 4 million tonnes of food a year or, on average, nearly 180 kilograms per person.

> **❶ Food waste is a global problem.**
> Annually, we waste an estimated
> **1.3 billion tonnes of food** worldwide.

Do you have any of the following problems?

- Lots of food is getting wasted in your home (**WARNING:** sad and soggy vegetables at the back of the crisper drawer).
- You never know what's for dinner (**WARNING:** when you get home from work, it's hard to come up with an idea for something to cook).
- You never have the right ingredients to cook what you want to eat (**WARNING:** you buy new food that goes into the fridge on top of the old).

Before you shop

Create a low-waste meal plan

One of the most effective ways to avoid waste and save money is taking a bit of time to plan your meals and your shopping trips *before* you go.

It means fewer trips to the supermarket and less impulse spending, or it may be frequent visits but only buying exactly what you need to keep it fresh.

You also need to factor in using leftovers more efficiently.

1. **Plan what you're going to eat over the next couple of days, and write a precise list of all the ingredients you'll need to make those meals.**

Don't be too ambitious and tell yourself you're going to cook every night, if you know you're not someone who loves cooking that much, or if you really don't have the time. (Also try your hand at cooking batches of food that you can eat over several meals, as long as you take care to store them properly and for not too long).

2. Check if you've already got some of the ingredients you need.

3. Go to the shops with your list, and buy only what's on it.

Go shopping in your own fridge first

Checking what you've already bought sounds simple, but we do forget. **Before you head to the shops, look at what lurks in the deep, dark depths at the back of the fridge.** Maybe you don't need another jar of pickles! If you're in a rush, just snap a pic.

Remember your reusable bags

Reusable bags are no good sitting at home or locked in your car. Read about all the problems with single-use plastic bags in Chapter 11.

1MW solutions

When we asked our community for their ideas on avoiding food waste, we literally received thousands of replies. So, here goes with some of my favourites:

We waste less food when I write on a whiteboard what produce is in the refrigerator.

I often make a pot of delicious 'fridge soup' with whatever is left in the crisper drawer at the end of the week, before I restock with fresh vegies.

I grow food from my food scraps. I have a mango tree, avocado tree and lemon tree – all grown from scraps.

Freeze grated cheese and crumbled stale bread for quick toppings.

Freeze peelings and scraps until you have enough to cook stock.

Invite a couple of friends over for a bulk cooking session. Each person nominates a favourite dish to cook and brings enough ingredients to serve about 20 people. At the end, divvy up the cooked food into reusable containers so that everyone goes home with a variety of ready-made meals to serve their family for the week ahead.

Don't put dressing on salad. Let people add their own when it's on their plate. Undressed salad keeps better (doesn't go soggy) and can be used the next day.

When my milk or cheese is about to turn, I make a heap of béchamel sauce and freeze it.

We eat mostly vegetarian food, so any inedible leftovers go into our worm farm and the castings are used to enrich our garden soil.

Muffins made with fruit pulp from a juicer.

As I don't have a compost bin in my block of flats, my food scraps have to wait a week before going to the community compost. I keep food scraps in a mushroom (paper) bag in the freezer. They stay frozen for the journey to the community garden, and the paper bag is also used in the composting process.

Blitz and freeze any soggy tomatoes to use later in pasta sauces.

I keep a shelf in my fridge that's just for leftovers or things that will spoil soon. Keeping those things right in front reminds us to use them first.

Freeze sliced banana for ice-creams and smoothies.

Bring children up with the expectation to finish what is on their plate.

If you don't have a compost bin, dig trenches in your garden beds and add fruit and vegie scraps to them and cover over. Add some mulch if you have some.

Wrap fresh loose-leaf produce (e.g., baby spinach, basil, rocket) in a damp cotton tea towel or muslin cloth to keep it fresh for longer.

Roughly once a week, I prepare 'Mum's crazy mixed-up dinner' to use things up. My kids might end up with ½ sausage, leftover veg, ⅓ apple, ⅓ bread roll, some leftover pasta, dip about to expire, etc. They love it!

Coffee grounds and ground-up eggshells are good for garden soil.

And here is mine. Have a tapas night once a week with little portions of lots of different foods – whatever is in the fridge. This gets most of the food eaten and, with what's left, we make omelettes the next morning.

Blitz any fruit that is about to go off and pour into ice-cube trays – a great snack for the kids.

Ferment, ferment, ferment. Pickle vegetables when there's a glut.

A word on best-before dates

Use-by dates should always be followed, but best-before dates are a suggestion. These dates are determined by the manufacturers; they often mark products as expiring sooner than they actually will. Many people throw away food that has passed these dates, even if only marginally, and it's actually still safe to eat.

You should use your senses – sight, smell and touch – to figure out if your food is spoiled, just like our grandparents did before there was any such thing as a best-before date.

Smell it
Has the odour changed from when you first purchased it?
Is there a whiff of 'not right'?

Look at it
Are there visible signs of decay, discolouration or mould?

Feel it
Is your meat slimy, or are your dairy products chunky?

At the store

Stick to your list

You've spent valuable time planning your meals for the week to reduce waste, so now's the moment to stay strong. When food shopping, **avoid the temptation to buy items you don't really need.**

Forgo the 2-for-1 deals, those prominently placed 'specials' at the end of the aisles.

Buy in season

Flying food halfway around the world so we can eat it out of season just doesn't make sense. It brings all those food miles with it. By law in Australia, supermarkets now need to say where fresh produce is from.

Wherever you live, choose produce grown in your country and sourced as locally as possible. That means that, by default, you'll be eating food that's in season.

Locally grown food is also often less expensive; it's going to last much longer in your fridge or fruit bowl because it hasn't already lived out its glory days in a warehouse or on the back of a truck. And, because food loses nutrients over time, what you eat will be a lot better for you – fresh really is best!

> **!** On one farm in Bundaberg, Queensland, only **13% of the tomatoes** grown meet the cosmetic standards for supermarkets. The remaining **87% are thrown away!**

Buy the 'ugly' fruit

Food waste isn't just happening in the kitchen bin. Traditionally, our supermarkets have been ditching up to 40 per cent of fruit and veg before it even hits the stores, because they don't pass the beauty test: it wasn't the right colour, was a bit lumpy or was a little too big or small.

Things are improving, at least in some stores. **Start selecting produce from the 'imperfect' sections that an increasing number of stores have created.**

Direct from the farmer

Another way of ensuring your food is local is by shopping from farmers' markets whenever possible. A lot of the time, **farmers will bring their imperfect produce to the markets as well, produce that wouldn't usually make it to supermarket shelves and would otherwise have been wasted.**

There is also an emotional connection when you meet the growers, and that, in itself, helps you to stop wasting the food they've grown.

And if there isn't a farmers' market near you, explore other options such as a fresh fruit and veg delivery service that delivers locally grown food. Or maybe there's a fresh food co-op close by.

Cut back on plastic wherever you can

Leave the overpackaged fruit and vegies behind. Buy loose produce instead.

See page 99 to read more about the 1 Million Women #Leaveitontheshelf campaign.

Fresh food starts with an organised fridge

Fridges aren't just giant cold cupboards, they're extremely intricate household appliances with many inbuilt ways to keep your food fresh. However, many of us don't use our fridge compartments effectively. **Storing food correctly and safely is an important step in minimising food wastage**, saving money and ensuring that our food nourishes our bodies.

If you're buying fruit and veg, and not cooking straight away, then diligent storage is vital. If you've never stored your washed baby spinach in an airtight container before, you will be impressed by how long it lasts in there. Even more than a week later, it'll still have crunch. Chopped celery can be stored in a similar way, with a little water in the container, or you can put the whole bunch in a glass of water to keep it from wilting. Herbs last a whole lot longer when wrapped in a damp tea towel if they are the hardy kind (such as rosemary) or in a jar if they're more delicate (such as coriander).

Note that some vegies, including potatoes and onions, are best stored in a cool, dark place but not in the fridge.

To start with, put all your newly purchased foods in the back of the fridge and shift the older foods to the front, so you can remember to eat them before they spoil.

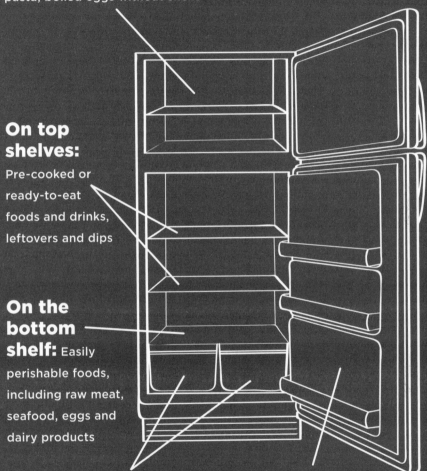

In the freezer: Meat, sauces, quality leftovers, freezable fruit, hummus, bread, ginger, cooked rice, pasta, boiled eggs without shells

On top shelves:
Pre-cooked or ready-to-eat foods and drinks, leftovers and dips

On the bottom shelf: Easily perishable foods, including raw meat, seafood, eggs and dairy products

In the crisper drawers:
Keep your fruit and vegetables separate to ensure they don't ripen too quickly.

In the door:
Non-perishable foods such as condiments and juices

Back at home

1. Cook your meals as soon as possible so you have fresh food ready to be eaten (and no ingredients are left wilting).

2. Keep lots of reusable airtight containers and jars handy for storing food, including leftovers.

Control serving sizes

Don't overfill plates, only to have to scrape that wasted food into one big dish at the end of a meal. **It's easier to come back for seconds than to put too much on the plates in the first place.**

Eat your leftovers

Plan your meals around your leftovers – a handy lunch for work or school the next day. Or maybe you can turn that huge bowl of leftover rice into fried rice, or that pasta into a yummy bake the next night.

Itsy bitsy recipes

Develop your own cuisine tricks for using up the bits and pieces of perishable foods that accumulate in the fridge: a splash of yoghurt here, a piece of cheese there, that last bit of salsa in the jar – it's all delicious if used and not wasted.

Scraps are food, too!

Start using entire vegetables, from root to stem. Add herbs stalks as well as the leaves to soups; collect carrot tops, onion skins and vegie scraps in the freezer for slow-cooked stock; save bones from a roast chook for broth; pop wilted vegies into a stir-fry or soup. Your potato peels are still full of nutrients and, cooked up, will make great chips.

Celebrate without waste

'You bring the salad, Mum makes dessert and I'm on mains.' Food is such a big part of special family occasions and celebrations, and it's important to not overdo the catering. **Plan extra carefully to get food portions and selections right, and have a strategy to use or distribute any leftovers.**

Compost

Composting is only for the food that you absolutely, definitely can't still eat – somehow. Note the use of the word 'food' here, because that's exactly what it still is! If people can't eat it, then as the last resort before composting, give this food to any animals you've got – chickens, or maybe your dog will like it – or, if you haven't got any of those, a worm farm is fantastic. **No matter what, we should make sure the last remaining nutrients in our food go somewhere**, whether it's back into the soil to feed future plants through compost, or into more mouths. If you can't compost all your food for whatever reason, see if you've got neighbours or relatives or friends nearby who'd like to add your food to their compost pile.

See page 208-9 for instructions to start and maintain your own compost bin.

Grow your own

Your own backyard is as local as you can get. **If you've got the sunshine and a bit of soil, give growing your own food a try.** Even if you don't think you have space, try growing vegies on the verge, or create a vertical garden made from recycled timber pallets on your verandah; or a few pots on your balcony.

When you've harvested a little extra, see if there's a produce swap nearby, where you can trade your excess cucumbers with someone who's got too many strawberries, or make pickles and preserves.

DIY: Grow your own lemon tree

Growing a lemon tree from seed is so satisfying. While it's not guaranteed to bear fruit, it's absolutely worth giving it a go.

You'll need:

- An organic lemon – often, non-organic lemons contain seeds that won't germinate
- Fertile potting soil
- Natural fertilisers, such as compost or worm castings
- A seedling pot
- A sunny, indoor growing location

Directions:

1. Fill the seedling pot to about 2 cm below the rim with soil. Moisten the soil so it's damp, but not soaked.

2. Take a seed from your lemon, and wipe away or suck off the pulp to clean the seed.

3. Plant the seed straight away, while it's still moist, about 1.5 cm deep in the centre of the pot. Cover with soil and gently moisten directly above the seed with water from a spray bottle.

4. Cover the pot with cling wrap, using a rubber band to seal the edges. Poke a few small holes in the cling wrap. Place the pot in a warm, sunny location. Make sure your growing lemon tree gets at least eight hours of sunlight a day; you can supplement with a grow light, if needed.

5. You don't want the soil to dry out, so spray with water every few days. Be careful to not overwater – you don't want the water to puddle.

6. In about two weeks, a little sprout should emerge. Remove the cling wrap, water occasionally to dampen the soil, and feed with some organic fertiliser.

7. Replant into a larger pot when needed, water regularly (note that a more mature plant won't need watering as regularly as the seedling did), and keep an eye out for pests and diseases.

DIY: Start a compost bin

Why compost?

Composting will reduce your overall food waste, and it will create a nutrient-rich soil commonly known as 'black gold' to use in your garden.

What to compost

Nitrogen rich (fresh greens)

- Grass
- Vegetable scraps
- Fruit scraps

Carbon rich (dry browns)

- Hay and straw
- Wood ash
- Pine needles
- Dead leaves
- Shredded paper
- Sawdust

How to layer compost: from bottom to top

1. Carbon (dry browns)
2. Water
3. Carbon (dry browns)
4. Nitrogen (fresh greens)
5. Carbon (dry browns)
6. Water
7. Nitrogen (fresh greens)
8. Carbon (dry browns)

Things to remember:

Whenever you add 'fresh greens' to your compost, make sure to add a pile of 'dry browns' over it to ensure it decomposes correctly.

Pile: Turn the pile every couple of weeks. It's important to regularly turn your compost heap to aerate it.

Bin: Turn the bin every couple of days.

Why does my compost smell?

Composts smell when they get too wet! This may be from rain or overwatering. To fix the problem, add more 'dry browns' to the pile, and turn it.

Will it compost?

YES: Fruit and vegie scraps, garden clippings (nothing too big, and nothing diseased), coffee grounds, tea leaves and bread, along with these unexpected items:

- Old toothpicks and golf tees (wood, not plastic)
- Ice-cream sticks (wood)
- Shoelaces (made from a natural fibre such as cotton, and remove any metal or plastic ends before they go in the compost)
- After-dinner plate scrapings – but not meat or dairy
- Old cardboard business cards
- Stale marshmallows, hard-as-a-rock jelly beans and other sweets
- Fabric scraps and thread snips from your latest craft project (natural fibres only: cotton, wool and felt)
- Toenail and fingernail clippings (eww, but true)
- Animal and human hair (eww, again – time to clean out your hairbrush)
- Pencil shavings
- Pulp from your juicer
- Dust bunnies (aka the fluff that gathers under your furniture)
- Dead insects (finally, somewhere to lay to rest those thousands of dead flies discovered in the attic)
- Rabbit, hamster, guinea pig and bird-cage clean-outs – but not dog or cat droppings
- Used matches
- Q-tips (but only wood)
- Lint
- Ropes (natural fibres only)
- Leather
- Fireplace ashes

NO:

- Dairy
- Meat
- Fish
- Tea bags (many types contain bits of plastic)
- Dog or cat droppings
- Citrus peels
- Onions (for worm farms)

OUR THANKS TO REBECCA RICHARDSON

DIY: Make your own worm farm

Each day, a composting worm will consume approximately half its body weight in food. This means you can feed your worm farm a few handfuls every few days. Cut food into small pieces – this makes it easier for the worms to eat.

Worm farms process less food than a compost bin, so make sure you're not overfeeding your worms. Uneaten food will begin to smell and attract unwanted pests. Once it's established and the worms are breeding, you can try feeding them more food.

Note: This DIY incorporates styrofoam boxes. You can get used ones from your local supermarket. If they are not salvaged for this second life, they will simply be broken down at the back of the supermarket and be reduced to a single-use life.

A MILLION THANKS TO THE GORGEOUS LISH FEYER FROM *GREEN IT YOURSELF* FOR THIS DIY WORM FARM.

You'll need:

- 2 styrofoam boxes with lids that are used to hold produce from a local supermarket (broccoli boxes are great for this). Grab your boxes before they get sent to landfill.
- 1 block of coir, a natural coconut fibre
- Approximately 1000 live compost worms. Source these from a garden centre or hardware store, or ask worm-farming friends whether they have any to share with you.
- A bucket
- An old towel, blanket or T-shirt

Directions:

1. Poke multiple holes along the bottom surface of one of the styrofoam boxes.

2. Mix the block (or brick) of coir with water and, once absorbed into the block so it's nice and moist, spread this along the bottom of your box as the base.

3. Place your worms on top of the coir base.

4. Place the worm-filled box on top of the empty second styrofoam box.

5. Poke a hole about 2.5 cm from the bottom of the empty (lower) box, which will drain the worm liquid (wee) that will filter from the top box into the bottom. Place a bucket under this hole to collect the worm liquid.

6. Give your worms a little bit of food to get started. You can feed them a bit more and a bit more and, as the worms start to multiply, you can increase the amount of food waste you feed them. Worms don't have teeth so if you give them a big chunk of potato, for example, it will take them some time to munch through it. So, it's always best to chop food up (or even blend it). The smaller the food pieces, the quicker they will get through it. Worms will eat almost anything you don't, but avoid feeding them citrus, acidic foods, onion, meat and dairy.

7. Place a damp old towel, blanket or T-shirt over the top of the worm box to keep it cool and dark, then place the lid on top. Worms don't like to be too wet, too dry or too hot, and the damp towel keeps them cool.

8. To make a rich plant fertiliser, dilute 1 part worm liquid with 7 parts water. Use it around your garden – it's incredible garden fertiliser.

Now, the story of meat

At least 15 per cent of our carbon emissions come from eating meat due to the fact that land is cleared to raise animals, especially cattle that are producing methane throughout their lives. Some cattle aren't just eating grass, once they move from paddock to feedlot, but rather grain that has been grown and shipped from elsewhere (more contribution to food miles).

'**Our appetite for meat is a major driver of climate change ... A shift to healthier patterns of meat-eating could bring a quarter of the emissions reductions we need**,' as stated in a 2015 report released by London-based Chatham House called *Changing Climate, Changing Diets: Pathways to Lower Meat Consumption*.

Meat consumption

Overall, livestock production accounts for more direct greenhouse gases than all trains, ships, planes and road transportation combined. Meat consumption is on track to rise by 75 per cent by 2050, compared with 40 per cent for cereals. By eating more plant-based foods, instead of meat, we can significantly reduce global emissions and have a huge positive impact on our planet.

Why do we love meat so much? Well, it's a big part of our culture and traditions for starters, and, in this world of plenty, it's more accessible than ever.

Which meat has the biggest impact?

Beef is, on average, around eight times more emissions-intensive on a per-kilo-of-product basis than chicken and over six times more emissions-intensive than pork, says the Chatham House Report.

Eat less meat – notably, less red meat

If you already don't eat meat, the planet thanks you.

If you're thinking about not eating meat, now is the time to do it. **Start by having one or two meat-free days each week, then aim to cut your meat consumption in half.** Reducing our meat consumption (but most notably our red meat consumption) will make a massive difference. And, when you do eat meat, find out exactly where it came from and how it was raised. Go to a local butcher, rather than the supermarket, and ask questions.

Many nutritionists say that it's better for you to get protein from vegetable sources such as chickpeas, nuts, lentils and beans, rather than red meat. These plant-based ingredients are often cheaper, and you can substitute them into most dishes that would usually contain meat. Doing this some of the time, even if not all the time, makes a difference.

Cook more vegetarian meals

Jump onto the internet, and search for the vegetarian version of anything you're thinking of making – there are plenty of recipes. You can also choose the vegetarian option when you go out to eat, which will provide some good inspiration.

If your fellow house dwellers are open to changing their cooking and eating habits, **try cutting down your meat consumption or going vegetarian as a whole household for a couple of weeks.** This will give everyone a taste of how delicious meat-free food can be. Then, at the end of the two weeks, have another chat and see if you want to continue like this. That way, it doesn't feel like such a huge commitment, and you'll have a chance to debrief on how it went.

Dairy is an issue, too

For the same reasons meat is a problem contributing to carbon emissions, so is dairy. Look for simple opportunities to reduce any excessive dairy consumption, including milk, cheese, yoghurt, ice-cream and milk chocolate.

- Change to drinking your tea and coffee without milk.
- Substitute cow's milk with soy milk or almond milk.
- Switch from eating cereal to toast for breakfast.
- Drink more water.
- Make ice-cream and milk chocolate occasional treats, rather than regular daily fare.

Quick ways to reduce food waste

DO TODAY:

☐ Cook produce from root to stem.

☐ Be mindful of serving sizes.

☐ Eat your leftovers.

☐ Move perishables on the verge of going off
to the front of the fridge, and eat before they spoil.

☐ Save food scraps for soup bases and stock,
or to compost.

☐ Check the fridge before you go shopping.

☐ Check use-by and best-before dates.

PLAN TO DO:

☐ Learn to store food properly.

☐ Buy 'imperfect' fruit and vegies.

☐ Leave overpackaged fruit and vegies on the shelf,
and buy loose products instead.

☐ Develop 'itsy bitsy' recipes using up small amounts
of perishable food and leftovers.

☐ Create a low-waste meal plan for the week.

☐ Grow your own.

☐ Shop at farmers' markets or food co-ops whenever
possible.

☐ Build a compost pile, start a worm farm or raise
some chickens.

Refuse

Reuse

Repair

Repurpose

Share

And **then** recycle

11

Plastic toolkit

A life without plastic

What does plastic have to do with climate change, anyway? **As it turns out, a lot.**

For starters, most plastic is made from oil or gas, fossil fuels that, in turn, use even more fossil fuels to be extracted from the ground. Then there's the manufacturing, transportation, use and disposal of the plastic to account for.

It's estimated that, **globally, we're now producing more than 80 million tonnes of plastic packaging annually. Alarmingly, just 5 per cent of this is reused.** A report by the UK's Ellen MacArthur Foundation estimates that, by 2050, making plastic could account for 20 per cent of total oil consumption and 15 per cent of the global carbon budget. (This is the budget to stick to if we, as a planet, aim to keep global temperatures from rising more than 2°C, and we should all pursue efforts to limit the temperature increase to just 1.5°C.)

Plastic clogs landfill, pollutes ecosystems, chokes the ocean and is even entering our food chain. The plastic industry also creates billions and billions of tonnes of carbon pollution every year. Plastic's killing our marine life, our birds, and it's being consumed by us via the fish we eat.

Clearly, we have a problem with plastic.

There's a garbage patch in the Pacific Ocean that's now three times the size of France, which contains approximately 1.8 trillion pieces of plastic weighing as much as 500 jumbo jets. It's not the only one in the world, but it's the biggest.

The reasons plastic became so popular in our society are the same reasons it's so damaging – it's durable, cheap to make and long-lasting. The low cost of manufacturing plastic has meant its production has increased rapidly since the 1940s. But it's longevity means that **nearly every single piece of discarded plastic ever made still exists on the planet today – even if it's breaking down into smaller pieces – and it will likely exist for hundreds of years to come.**

I know these stats sound overwhelming, and you might be thinking: *What can I possibly do to change this?* But I know this story is already changing.

In 2006, when I was having my epiphany about climate change, you couldn't watch the news or current affairs, open a newspaper or listen to the radio without there being something on climate change. I think the media played an enormous role in bringing climate change to our attention back then, and making the issues accessible for everyone to understand. The same thing is happening right now with the story of plastic (and waste).

We all have a huge role to play in this story, and we can start right now by simply saying NO to plastic in our life wherever we can.

Your plastic-free journey

Start your plastic-free journey by, firstly, looking at what you do and what you buy throughout an average day/week/month. Take note of what is usually wrapped or contained in plastic.

Assess how you can switch to reusable alternatives, such as airtight containers and your own shopping bags. Keep your plastic alternatives either in your bag, in your car or somewhere close by so you won't forget them.

Say goodbye to single-use plastics forever

Those lunchtime takeaway boxes, that smoothie on the run, the quick coffee: you use these containers for minutes then into the landfill they go for the next few hundred years.

> ❗ In fact, **half of all the plastic** we use, we only **use once.**

Remember, recycling is not an excuse for buying more plastic

Just because it's recyclable doesn't mean we should buy it. Everything we buy comes with the embodied energy needed to make it. **Avoiding something packaged in plastic must be the first step.**

So, every time you pick up something packaged in plastic, ask yourself:

- Do I really need this?
- Can I make it myself?
- Could I get this in my own container?
- What can I do with this plastic when I'm finished? (e.g., the hummus dip, the plastic container of olives, the strawberry punnets, the prepacked cheeses)

If we are to be serious about cutting back on plastic, we need to look at it all. **Leaving plastic on the shelf is the most powerful and direct message we can send** back to the store owners and manufacturers that this needs to change.

It wasn't all that long ago when plastic hadn't yet been inserted into every aspect of our lives. Bread used to come in paper bags. Milk was sold in glass bottles.

Here's a revealing statistic: of the estimated 8300 million metric tonnes (picture 1537 million Asian elephants, if that helps) of plastic produced in the last 50 years, half of this was produced only in the past 13 years.

Just say NO. It's that simple.
Simply refuse to buy it or be given it.
It's so liberating when you do.

Here are 12 easy ways to get you started.
Stop using:

- Plastic bags
- Cling wrap
- Plastic straws
- Packets of chips
- Soy sauce fish packets
- Plastic cutlery
- Coffee stirrers
- Shiny wrapping paper (it's probably coated in plastic)
- Plastic party cups
- Disposable plastic plates
- Takeaway plastic anything
- Overpackaged plastic everything!

And a bonus one:

- Drink loose-leaf tea – a lot of teabags have a thin plastic coating.

How bad is the plastic-bag problem?

Worldwide, 1 trillion plastic bags are estimated to be used and discarded every year. **That's 2 million every 60 seconds.**

In Australia alone, we use 3.92 billion plastic bags *each year*. **That's about 170 plastic bags per person, per year!**

❗ It is estimated that **8.7 plastic supermarket bags** contain enough embodied petroleum energy to **drive a car for 1 kilometre**.

❗ In China, **3 billion** single-use plastic bags are used **every day**.

Do you get caught without a reusable bag?

Have a think about your roadblocks: do you need to keep bags in your car, or a little one that can pack up small in your handbag?

Speak up: Get in quickly when you're checking out, and let the shop assistant know you don't need a bag. Sometimes, they place your goods in a bag before you've had time to think.

Share bags: Keep a couple of reusable bags at your workplace to share for those lunchtime dashes to the shops.

And for those small, plastic, produce bags: Consider whether you need any small bag at all. If you really do, either use a paper bag (the stores have them for mushrooms) or reuse your own from home.

✅ During the first six months after the **UK started charging for single-use plastic bags** in 2015, the number of bags handed out **dropped from 7.6 billion to about 500 million.** Now that's progress!

DIY: Make a tote bag from an old T-shirt

You'll need:

- An old T-shirt
- Scissors

Directions:

1. Cut along the T-shirt's neckline.
2. Cut on the inside of the sleeve's shoulder seam.
3. Repeat on the other sleeve.
4. Turn your T-shirt inside out.
5. At 1 cm intervals, cut slits about 2.5 cm long along the hemline.
6. With the T-shirt lying flat, knot the front and back strips of material together.
7. Tie neighbouring knots together.
8. Turn right side out, and you're ready to use.

Micro-plastics, microbeads and microfibres

Made from the same plastic used to manufacture plastic bottles, micro-plastics are tiny pieces of plastic measuring less than 5 mm. They come from a variety of sources, including from larger plastic debris that has degraded into smaller and smaller pieces.

Microbeads are a form of micro-plastics. These are tiny particles of plastic, measuring between 0.1 and 0.5 mm, which are found in personal-care products such as facial scrubs, body peels and shampoos, and can also be used to add colour to toothpaste. These, along with synthetic clothing fibres, cause great environmental damage when they're flushed through our waterways to be ingested by marine life. Read more about these harmful plastics and how to avoid them in Chapter 12.

! **Synthetic clothing** can shed **700,000 microfibres with every wash**. Washing our clothes less makes a difference.

! Glitter is litter! Most glitter is made from polyethylene, the same substance found in plastic bags.

Stop using glitter altogether, or choose products that get their **sparkle from mica or minerals**.

The essentials to going plastic-free

Reusable shopping bags

You have the power to cut plastic-bag use right now. All you need to do is use reusable bags whenever you shop. And there is good news on this front. Progress is being made. Many supermarkets around the world, including in Australia, have now banished single-use plastic bags altogether.

And that's not because of some high-minded ideal on their part – it's good business. It's the result of enough people like you and me changing our personal shopping habits, and having our choices influence demand and thus supply.

A lot of retailers offer compostable and biodegradable bags, but are they any better? The short answer is no.

Here is a quick rundown.

Compostable bags

The word 'compostable' is misleading to the average consumer. You'd think a bag labelled 'compostable' means you could throw it in your backyard compost alongside your fruit and vegie scraps, right? **Wrong.**

Compostable bags can biodegrade, but only under certain conditions – they need to be composted in a specific composting facility (of which there are very few in Australia).

Biodegradable plastic bags

Biodegradable bags are made from plant-based materials such as corn and wheat starch rather than petroleum.

However, when it comes to this kind of plastic, there are certain conditions required for the bag to begin to biodegrade:

- Firstly, temperatures need to reach 50°C.
- Secondly, the bag needs to be exposed to UV light.
- Plus, if biodegradable bags are sent to landfill, they break down just like food waste to produce methane, a greenhouse gas that is 23 times more powerful than carbon dioxide.

Bin liners?

But what about using single-use plastic bags for your bin, I hear you wonder? Don't need them! **Your bin can go naked.**

Or, if you really need something in your bin to contain drippy rubbish, use newspaper on the bottom or salvage packaging from something else. Reusing the bag from inside a cereal box is a good solution.

A water bottle

Astoundingly, it takes 3–7 litres of water and 1 litre of oil to produce a single litre of bottled water. And that doesn't even begin to cover the **greenhouse gases produced by pumping said water out of the ground, before packaging, transporting and chilling it**.

While many plastic water bottles are recyclable, most of them end up in landfill, where **they could take an incredible 1000 years to break down**. Or worse, they break up into micro-plastic particles in the ocean and kill marine life.

These days, we're never too far from a water source, so it's easy to refill whenever necessary. Upcycle a sturdy bottle from home, or you could purchase a stainless-steel water bottle – they are long-lasting and, although they cost more money upfront, they'll save you money in the long run.

> **!** In London, the average person uses **175 plastic water bottles** each year.

Glass drink containers

Whenever possible, purchase glass instead of plastic milk containers and drink bottles.

Also, check whether your local area has the facilities to recycle wax-coated cardboard milk and beverage cartons.

A coffee cup

Many paper coffee cups are lined with waterproof plastic and can't be recycled. If they are placed in with regular paper and cardboard recycling, they could contaminate the entire batch, causing the lot to be dumped in landfill.

If we all switched to reusable cups, we would stop 500 billion disposable coffee cups and their lids going to landfill every year. Get yourself a reusable cup, or you could even use an old heatproof jar for your takeaway coffee or tea.

Make a pact with yourself right now that if you forget your cup you don't get a takeaway coffee – either drink it at the café in their ceramic cup or go without.

❗ If we used **refillable coffee pods** instead of disposable, we would **divert 20 billion coffee pods from landfills** every year.

Coffee pods

While we're talking coffee, please say no to coffee pods. **A staggering 55 million pods get thrown away every single day, and one coffee pod takes 500 years to break down.** Thanks a bunch, George Clooney!

It's not good enough to say that some get recycled. Not enough of them do to make an impact. The 6 grams of coffee in the pod comes with 3 grams of environmentally damaging packaging. Why not just make coffee in a plunger or stovetop coffee pot instead?

Take your own container

Ask whether you can fill your own clean containers at the deli or takeaway shop. In Australia, there is no law against it. If the deli or takeaway staff say no, politely inform them that there are no regulations against it, or ask them to check with a manager. Don't take no for an answer if you can help it.

Some businesses may create their own policy, saying they won't allow you to bring your own container, but there are no legal grounds for that. You can ask them why they've chosen to make environmentally irresponsible business decisions (saying this publicly, on social media, can often bring their attention to the matter faster).

Plastic cutlery

Keep a set of reusable chopsticks and metal cutlery in your bag for when you leave the house. Wrap them up in a cloth napkin to also avoid single-use paper hand towels.

Cut out cling wrap

Cling wrap is one of those plastics that are rarely ever recycled. It can't go into normal household plastic recycling bins, as it melts at a different temperature to the other plastics and causes problems with the machines.

Use alternatives such as beeswax wraps (see pages 236–7 for how to make your own), reusable containers, cover bowls with a tight-fitting plate – or don't wrap the food item at all (but use quickly, before it spoils).

Say no to straws

Every day, 6 billion plastic straws are used worldwide, and one straw can take 200 years to break down.

For most of us, they're one of the most redundant pieces of plastic around. Say no to straws at cafes and bars.

Or get yourself a reusable metal or bamboo straw if you need to keep a straw with you to use when you're out.

Buy a bamboo toothbrush

In Australia alone, 30 million plastic toothbrushes are thrown away every year. Switching to bamboo toothbrushes will keep that plastic out of landfills.

Use one stainless-steel razor instead of a pack of plastic throwaways

In a lifetime, people use hundreds of disposable razors. These will exist for centuries. Change to stainless steel or an electric razor that can be reused indefinitely.

Planet-friendly kitchen toolkit

- [ ] Bicarbonate of soda: grime buster and great for just about everything
- [ ] Clove oil: mould remover
- [ ] Compostable scourer
- [ ] Food-grade soap bar
- [ ] Glass jars
- [ ] Reusable cloth: for wiping benches and dishes
- [ ] Beeswax food wraps
- [ ] White vinegar: surface wipe and disinfectant
- [ ] Wooden and metal utensils

Plastic-free bathroom toolkit

- [ ] Bamboo toothbrush
- [ ] Biodegradable bath brush
- [ ] Conditioner bar
- [ ] Lotion bar
- [ ] Shampoo bar
- [ ] Soap bar
- [ ] Paper-wrapped toilet paper
- [ ] Stainless-steel razor

DIY: Reusable beeswax wraps

If you want to kick the single-use disposable habit, then homemade beeswax wraps are a great alternative to plastic cling wrap.

Use them to cover food and keep it fresh (as you would cling wrap) but, **importantly, we recommend that you don't use beeswax wraps on dairy or meat products as they may leave harmful bacteria on the wrap.**

Beeswax wraps are reusable and can easily be refreshed if the wax starts to wear. If you notice that the wax is wearing away after a few months, you can simply sprinkle on another layer and repeat the ironing process to make them brand new again!

How do I clean them?

Wash your wraps in cold or lukewarm water with a mild soap.

You'll need:

- 100% organic cotton fabric squares (cut a few different sizes)
- About 1 cup of grated beeswax
- An iron
- An old towel or blanket that you don't mind getting waxy
- A few sheets of baking paper

Directions:

1. Lay down your towel, and turn on the iron to heat it up.
2. Place a few sheets of greaseproof paper onto the towel, then lay one of your cotton squares on top of this.
3. Evenly sprinkle beeswax over the cotton square, ensuring that you go right to the edges.
4. Place a few more sheets of greaseproof paper on top of the wax and cotton square, then iron over it.
5. Peel back the paper to check that the wax is evenly melted. You may need to sprinkle more on for even coverage, then replace the paper and iron again.
6. Peel the wax-infused cotton square off the greaseproof paper and let it cool. Then set it on a wire rack or peg it onto a line of string to dry.
7. Use as you would cling wrap, with the exception of covering dairy or meat products.

12

Fashion and cosmetics toolkit

Wear it well

True fashion is delightful – it's an artform. But fast fashion is something else entirely. I'm talking about the stores that churn and burn through trends, offering up-to-the-minute clothes with a spend less, buy more approach.

Before my epiphany, I would buy clothes on impulse, only to end up not liking what I'd bought. I'd buy cheap, specials (simply because they were on special) and always new.

I'm embarrassed to admit that, back then, if a shoe broke or a heel fell off, I'd buy another pair rather than have it repaired. *They weren't expensive*, I thought, *so it doesn't really matter.*

Not anymore! **Our clothes should never be disposable!**

They can be reused, mended, passed along to another person or recycled.

If we include the cosmetics industry in all of this – with all the packaging, chemicals and waste – we see the full effect of commercial exploitation, playing on our urge to *beautify* ourselves.

How did it get so bad?

The world now consumes a staggering 80 billion pieces of clothing each year. This is up 400 per cent from just two decades ago, yet we are now spending less because our clothes are a lot cheaper than they were.

The fast-fashion industry has exploded, with **some brands having 52 fashion seasons every year**, while others boast 400 micro-fashion cycles per month. All capitalising on the social pressure to always look on-trend, urging you to buy, buy, buy. Fast-fashion manufacturers have made clothes so cheap because they want you to buy today, chuck out tomorrow and buy again the next day.

It's estimated that more than half of the fast-fashion items produced are disposed of within less than a year.

The fashion industry is one of the world's largest polluters – from the pesticides used to grow textile crops to the toxic chemicals and dyes poisoning water supplies, and from the manufacturing waste to the carbon footprint of transporting garments globally and the discarded clothing itself.

Even if we donate our used clothes to second-hand shops when we're finished with them, only 10–15 per cent will actually end up being bought by someone. Because of poor quality, much will be sold off as industrial rags, and a lot will go to landfill.

> **(!) Less than 1% of material** used to produce clothing is **recycled into new clothes.**

How we can change the story

Fashion can be toxic – but the good news is that we can still enjoy it and help change the story at the same time. Our dollars, and how we spend them, send a clear signal: whether we accept the status quo, or whether we're backing people who are doing it better.

If we value the things we purchase, we take action to move away from that throwaway mentality.

We can do this if we follow a few key points, as summed up by the wonderful Vivienne Westwood:
'Buy less, choose well, make it last.'

Buy less

Most of us are guilty of buying more than we need. We must learn to buy less and love what we already have.

Once we get our head around living with fewer clothes, we'll be better placed to resist advertising that encourages us to constantly update.

- **Don't buy on impulse or shop when you're in a hurry.**
- **Don't buy things on special, just because they are on special.**
- **Take a breath before you buy something.**

Here's a good exercise.

Challenge yourself: don't buy any clothes for three months. If you see something you really, desperately want, tell yourself you can get it at the end of the three months. (If the item is no longer in store, you'll probably be able to find it online, or second-hand.)

The thing is, though, after those three months, you'll probably find you don't really want it anymore.

Go shopping in your own wardrobe

You'll be amazed at what you might find. To give clothes a new lease on life, rediscover what's crammed in the back of your closet and start wearing them again. Mix and match garments to create new styles, no matter how long you've had them for.

Apparently, **80 per cent of the time, most of us only wear 20 per cent of our clothes** – and the other 80 per cent of items just hangs around, looking pretty but unused.

Swap outfits

After you've gone digging in your own closet, suggest doing a clothing swap with a similarly sized friend. You may both find a perfect 'new' piece that's been languishing, unloved, in another closet. **One woman's trash can be another woman's treasure.**

Party in style

If you're going to a special event, instead of buying something you'll probably only ever wear once or twice, **consider renting a dress or borrowing one from a friend.**

Choose well

Ask yourself these questions:
- Is it made well?
- What's it made from?
- Who made it?
- Do I love it?

Choose quality over quantity

You might need to pay a bit more for something made well, but it'll be worth it in the long run, and should be doable if you're buying a lot less than you were before. Clothes that are made well will be manufactured from higher quality fabrics, so not only will you be able to wear them for longer, but the fabric won't wear as fast.

Pay attention to clothing labels

All synthetic clothing is a significant contributor to marine plastic pollution, mainly through microfibres going down the drain when we wash. Synthetic fabrics, most often found in polar fleece garments, leggings and athletic wear, are largely to blame for the release of these plastic fibres into waterways. Although natural fabrics also shed microfibres, these break down much quicker than their synthetic counterparts, which can take centuries to degrade.

Be selective

Some natural fibres use a lot less water to produce than cotton, for example, bamboo, linen, hemp and Tencel. Organic cotton is grown without pesticides and using recycled water systems, so is much better than conventionally produced cotton. Wool and silk are also good natural fibres with great properties, such as being anti-microbial (so they won't smell after you wash them), and they'll naturally keep you warmer when you need it.

Choose second-hand

Quality clothes made from natural fabrics are often expensive; it's hard to have a wardrobe full of pieces like this. For the sake of your wallet, and the planet, check out the second-hand stores first. Arm yourself with a list of natural fibres to look out for and a good idea of exactly what you want, then go hunting.

Instead of buying a brand-new polyester-cotton blend jumper that's hanging together by a couple of threads, for the same price, find yourself a pre-loved 100 per cent wool jumper that's going to last much longer.

Go natural

Next time you shop, choose natural fibres, which will eventually break down completely.

Bamboo

Cashmere

Cotton, particularly organic

Hemp

Jute

Linen

Silk

Tencel

Wool

Avoid synthetic fibres, which can take up to **200 years to break down, including:**

Nylon

Polyester

Rayon

Spandex/Lycra/elastane

Don't buy based on trends

Buy based on what makes you feel and look great. If you find yourself hesitating, that feeling probably won't go away, and the piece of clothing will just sit around gathering dust at the back of your wardrobe.

Can you pair it with at least three items in your wardrobe?

If not, put it back and keep looking. You don't need to own something that may have only a few wears in it.

Donate clothes

Identify which of your unwanted clothes are still in good condition but just don't fit, don't suit your style or you don't love anymore, and take them directly to shelters for those living rough on the streets or to organisations geared towards outfitting women re-entering the workforce. For everything else, give to charities.

Where was it made?

If a garment was made halfway across the world, the carbon miles it took to end up in your wardrobe have contributed to climate change.

If you buy online from overseas sources, by the time it arrives, your package may have travelled tens of thousands of kilometres, adding a big whack of CO_2 to the environment, especially if you select the airfreight delivery option to have it arrive within days.

Be wary of 'greenwashing'

Be on the lookout for those companies using marketing hype to falsely claim that their methods help the environment. Or those that have one product that is eco-friendly, while all their other products are unsustainable.

Hold brands accountable

Research from Oxfam highlights that **83 per cent of Australians surveyed want clothing retailers to stop being so secretive.** As consumers, we can pressure textile brands to review the impact of their fabrics; demand greater transparency in labelling; and invest in ethical clothing that covers a range of issues, including working conditions, exploitation, fair trade, sustainable production, environmental impacts and animal welfare.

We also need to hold brands accountable for how they manage unsold stocks. I know you will find this hard to believe, but many fashion brands simply burn their unsold stock because they don't want the clothes flooding the market at cheaper prices.

At 1 Million Women, we have launched a campaign asking big fashion brands such as Burberry – that burned £28.6 million of unsold clothing, accessories and perfume in 2017 – to publicly say they will no longer continue this disgraceful practice. In a world bursting with imaginative ideas about how to solve the climate crisis and deal with waste, it's astounding that any brand would set excess stock on fire. What a missed opportunity for one of these top brands to step up and become a leader in the fashion industry and repurpose their incredible fabrics and other unsold goods into something new instead.

#FashionNotFlames

#StopFashionBurning

> ❗ Up to **one-third** of all clothing produced across the planet annually **never gets sold.**

Make it last

Essentially, to make our precious clothes last longer, we need to wash them less. That might sound gross, but frequent washing is just another process we've been conditioned to expect.

Half the damage our clothes impart comes from when we wash them. They deteriorate faster, and older clothes that have been washed more will shed more microfibres. You might notice some brands of jeans recommend washing them as little as possible.

Keep it clean

When you cook or do messy tasks, put on an apron or wear some work clothes meant for just around the house, ones you don't mind if they have a few stains on them. Being daggy around the home is one of life's great pleasures.

Spot-clean

When you spill something on your clothes, try to spot-clean them straight away. Remember: a splash of soda water can work wonders! If you don't get a chance to treat a stain until later, still try to only clean that one area first, rather than the whole garment.

! More than **70% of the negative environmental footprint** associated with our clothing – such as **energy use, water consumption and pollution** – comes from how we care for our clothes, especially **doing laundry, drying and dry-cleaning.**

Hang smelly clothes in the sun

The UV rays in sunlight will help to kill any lingering bacteria that can cause clothes to smell. And if there's a nice breeze, that's even better as it will loosen the odour-causing particles. If you don't have an outside clothes line, hang the item next to an open window for at least an hour.

Mend or repurpose worn clothing

And if anything does happen to your clothes, repair them! If they're beyond repair, there's probably a way to upcycle them – remove old zippers to be used elsewhere, turn old T-shirts into bags (see page 225 for instructions), or even make T-shirt yarn. Having basic sewing skills under your belt is highly beneficial to the planet and your wallet.

Sunshine is 100% free

Whenever possible, hang clothes outside to dry in the sun. Get an airing rack to use indoors or under cover on rainy days. Your clothes will thank you – they will last longer than if they were getting bundled into the dryer.

> **!** If 1 million of us **ditched the clothes dryer**, we would save the energy equivalent of **taking 95,000 cars off the road.**

Wash in cold water

Using cold water saves both money and energy with every wash. Cold water isn't as harsh as warm or hot, helping to preserve clothes. Not to mention that **up to 90 per cent of the energy used by washing machines is to heat the water.**

Switch to soap nuts

Get some soap nuts to replace commercial laundry detergent. To use, simply place three or four soap nuts in a cloth bag, tie it up and throw it into the machine with your washing. **They are grey-water safe, completely biodegradable and gentle on sensitive skin.**

Avoid dry cleaning at all costs

The chemicals used in the dry-cleaning process are extremely damaging to the environment.

Eco-fashion ideas from our 1MW community

Buy a sewing machine. Clothes can be redesigned with old buttons, pieces of material, beads, zippers, etc.

If any of your friends or family are emptying out their wardrobes, check out items they don't want. You can find some good treasures!

To make my wardrobe more exciting, I rotate clothes in and out of storage regularly so I can get excited about them again.

I tend to cut off or add bits, resewing when needed. Most of my clothes make good cleaning cloths when I'm finished with them.

Two words – colourful wraps.

I much prefer to buy clothes made of natural fibres so I can compost them when they're old, rather than send them to landfill.

And mine – I think one of the most liberating things we can do is to find our own style, which outlasts trends. When I was growing up in the '70s, my mum was the most stylish person I knew. Her big beehive hairdo and her incredible dress sense were all hers. She wore her clothes year after year, and she never went out of fashion, because she owned her own look. (Still does. Love you, Mum xxx)

Build a capsule wardrobe

Rejecting the world of rampant overconsumption is a fabulous step towards living a low-carbon, low-waste life. Maintaining a capsule wardrobe reduces your impact on the planet and declutters your mind, making room for the more important things.

The term 'capsule wardrobe' was coined by British fashion icon Susie Faux in the 1970s. It consists of owning only 30 much-loved quality items that you wear exclusively. In order to get the most out of your capsule wardrobe, pick the necessities first and ensure all items have the versatility to be mixed and matched together.

Step 1: Categorise

Think practicality first. Create a category list of vital item types (excluding underwear and socks) before you start sorting.

What kinds of items do you wear most often? Dresses? Pants and T-shirts?

Depending on the weather in your part of the world, and your personal style, a good start is to add jeans, T-shirts, jackets, dresses, skirts, heels, flats, joggers, jewellery and sunglasses to this list.

Step 2: Sort

Dump every item from your wardrobe onto your bed or floor, and sort them into three different piles: keep, maybe and go.

Depending on how many items you started with (and how decisive you are!), it might take a while to whittle down the 'keep' and 'maybe' piles. Ask a friend to help you – someone who isn't emotionally attached to the clothing and gives great feedback.

Ask yourself, 'Have I worn this item in the last year?' If the answer is no, chances are you're not going to wear it again.

Sort until your 'keep' pile contains just 30 items, and make sure this pile contains the vital item types you wrote down in Step 1.

Then, give yourself a big pat on the back.

Step 3: Say goodbye

It's time to say goodbye to your unwanted clothes: upcycle, swap, donate or sell them at the markets.

If you have clothes that are very worn, they may not be able to be reused or resold at a charity store, so consider composting them if the fabric is biodegradable.

MANY THANKS TO TESSA MANARO

Cosmetics

True beauty comes from within

No amount of product can deliver this kind of exquisite beauty, and we should always be selective with what we put on our skin.

What can we do?

Just buy less. We don't need three-quarters of the beauty products we're told we need. Instead, we should eat well, sleep more, stress less and find contentment from within, then marry all that up with some basic good skincare routines.

When we do buy products, let's support beauty companies that are making every effort to simplify their packaging and products. Research your favourite brands and ensure they are doing their bit, at a minimum, to cut down on packaging and assist you in recycling.

There are so many brands now making their products in solid form, including body and skincare bars, shampoo and conditioner bars and shaving soap, thus negating not only the harsh preservatives necessary for liquid products but also the massive amount of packaging and plastic that comes in the form of bottles, caps and pourers.

Boycott microbeads – a totally useless ingredient!

Microbeads are minute pieces of plastic (defined as 0.1–0.5 mm in diameter) are used to give beauty products such as facial and body scrubs a grainy texture for exfoliation. **They're an ecological nightmare hastening an already dire problem – the global pollution of our waterways and oceans with plastic.**

Too small to be filtered out by sewerage systems, microbeads flow swiftly into our oceans, rivers and lakes, where they float and absorb toxins then are ingested by marine life.

Boycott buying any beauty products that contain microbeads. If you have products in your house containing microbeads, send them back to their manufacturers with a letter explaining why.

As consumers, we have the power to say, 'No, no way. I'm not having it' to these items. Be vigilant when shopping for cosmetics, and make sure these products are left on the shelf to gather dust.

> ❗ One tub of facial scrub can contain more than **300,000 microbeads!**

Label alert

Watch out for these ingredients in beauty products that contain microbeads:

- Polyethylene terephthalate (PET)
- Polylactic acid
- ADA 2014
- Polymethyl methacrylate (PMMA)
- Polypropylene (PP)
- Polyethylene (PE)
- Nylon

Make your own

Switch from store-bought cosmetics and toiletries to homemade. There are so many recipes online for personal-care products you can make from natural, easy-to-find ingredients.

While it may take a few tries to find the right recipe for your needs, ultimately, you'll be spending less money on beauty products. You'll know exactly what goes on your skin and can reuse your own glass jars to eliminate waste.

My favourite recipe for dry skin

Grab yourself a 100 ml bottle of rosehip oil, plus a small bottle of pure rose essential oil. Place 15 drops of rosehip oil in the palm of your hand, add 1 drop of rose essential oil, mix together well and massage over your skin morning and night.

Another lovely essential oil to try instead of rose oil is calendula oil, or you use 1 drop of each if you have both.

Make sure the oils you purchase are top quality.

NOTE:
Never put essential oils directly on your skin. Always remember to mix your essential oil in a carrier oil as they are too strong on their own. Use the ratio of 1 drop of essential oil to 15 drops of carrier oil. Usually, though, rose essential oil may already be mixed with a carrier oil, because its expensive.

DIY: Aloe vera gel mask

Aloe vera is known for its healing properties, which makes it the perfect mask for dry, sunburned or wind-burned skin.

You'll need:

- ½ overripe avocado (save the rest for dip!)
- 2 tablespoons aloe vera gel
- 1 tablespoon honey, warmed slightly to make it runny
- 1 teaspoon almond or coconut oil, plus extra for moisturising
- 1½ tablespoons rolled oats (not instant oatmeal)

Directions:

1. Place the avocado in a bowl, and mash with a fork.

2. Add the rest of the ingredients, and mix well.

3. Apply mask to freshly washed skin, and let it sit for 15 minutes. Rinse off with lukewarm water and a washcloth.

4. Then moisturise your skin with a dab more almond or coconut oil.

DIY: Coconut and lemon lip balm

Lip balms and lip glosses are popular hoarding items. Give this recipe a try before buying another one.

You'll need:

- 1 tablespoon hard wax (candelilla, soy or beeswax*)
- 2 tablespoons sunflower oil
- 1 tablespoon coconut oil
- 10 drops lemon essential oil
- A small saucepan
- A large saucepan
- A whisk or spoon (whisk is better)
- Glass droppers
- Clean containers, preferably reusing something you already own

Directions:

1. Finely chop or grate the wax, and place it in the small saucepan. Add the sunflower and coconut oils.

2. Heat approximately 2 cups water in the large saucepan over medium–high heat. Carefully place the wax-filled saucepan in the larger saucepan, making sure no water spills onto the wax.

3. Still over medium–high heat, gently whisk or stir the wax-and-oil mixture until melted.

4. Once melted, remove the saucepans from the heat and set aside.

5. Stir in the lemon essential oil.

6. Use the dropper to move the liquid balm to the empty containers. Let cool overnight – around 10 hours is best – before closing them.

NOTES:
*Importantly, people allergic to bee stings should not use beeswax.

Never put essential oils directly on your skin. Always remember to mix your essential oil in a carrier oil as they are too strong on their own. Use the ratio of 1 drop of essential oil to 15 drops of carrier oil.

Ending waste, period.

The average menstruation span is roughly 40 years per woman and, on average, we women may use up to 20 tampons a cycle.

20		**240**		**9600**
tampons a cycle	**=**	**tampons a year**	**=**	**tampons in a menstrual life cycle**

- 20 billion used tampons and pads are being dumped into landfills each year.
- Conventional sanitary pads can contain plastic that's equivalent to about four plastic bags, and the polyethylene plastic in pads can take about 500 years to decompose. Our periods are literally outliving us!

Have you ever considered using sanitary products that are safe for the environment, including reusable silicone menstrual cups, reusable cloth pads or menstruation underwear (aka period panties)? These products are easy to care for as well; simply rinse out cloth pads and underwear before washing as normal. Silicone cups are even easier – rinse out in warm/hot water every 12 hours.

❗ If 1 million of us switched to menstrual cups, we'd divert 240 million tampons (and their packaging) from going to landfill every year.

13

Overconsumption toolkit

Do you really need it?

Have you ever done a cull of your wardrobe, your junk drawer or even your whole house? How good is that feeling of freedom and lightness?

Living with less stuff means feeling like that … all the time.

Lots of us have a love–hate relationship with stuff. We want it the moment we need it but, when it's cluttering our homes, we don't want it any longer. And it's not only things we're consuming more and more of; it's the energy we use, the waste we leave behind us, and the whole bundle of impact that we have on our locale and the planet.

If the thought of going on a full-scale decluttering mission is daunting, you can try to build momentum with an easy win or two first.

Take a look around your house. **If you haven't used something in six months, then chances are, you never will.** These are the first items to target on your way to downsizing your life.

But before you decide to throw something out, consider how you can help the sharing economy by renting it out, giving it to someone else or lending it for a time.

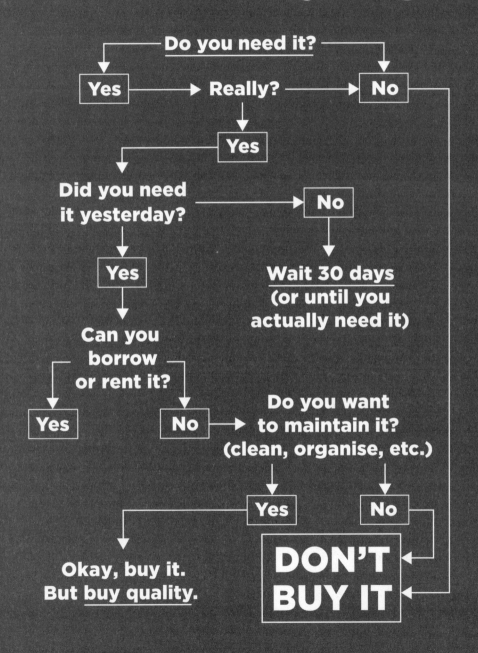

Let's embrace the sharing economy

Sometimes, there are things we *do* need – a gorgeous dress to wear at a wedding, or a drill for some key handywoman jobs around the home, or a toy for a one-year-old (who'll grow out of it in a year).

But do we really need to own these things ourselves?

This is where the sharing economy comes in. **We borrow or rent whatever we need for as long as we need it, then we give it back.** It means we have less clutter in our homes, we spend less money and we throw less away. And the sharing economy is growing faster than ever – **at the time of writing, the sharing industry will be worth US$335 billion by 2025.**

Let people know you've got something to share online – search for Buy, Swap and Sell groups – or find an app to help you.

Cars

Car-sharing is an awesome way to take care of incidental driving needs, particularly if you live close to public transport and can walk or ride a bike to work. There are so many companies (search the internet for them) that, once you subscribe, allow you to **rent cars as you need, for just an hour, a day or longer.**

If you want to go on a longer road trip, hiring a car could look expensive in the short term but, in the long term, you're the one saving money. And it means there's one less car in the world, either being used when public transport or walking was a better option, or not being used at all. See Chapter 15: Travel toolkit, to read more about getting around.

Toys

They're so fun that it's easy to go completely over the top. Who doesn't love to see kids having a great time? The problem is, however, **toys are often made from plastic, packaged in plastic, and (thanks to overenthusiastic play) liable to break and be irreparable.** There's no need to ban toys from the house, though. Just search online for 'toy library association' to find where you can borrow durable, quality toys, plus books and sports equipment at some locations. While often geared towards younger ages, most toy libraries will have items covering a range of educational and developmental stages.

Baby clothes

Babies grow. Fast. If you don't have someone to whom you can pass along outgrown – and possibly never worn! – clothes, or you don't have access to hand-me-downs, then outfitting infants and toddlers can be a huge waste of resources and money.

To reduce the financial burden and environmental impact, parents around the globe have banded together to form baby clothes–sharing networks. Search the internet or ask other new parents to see what's available in your area.

Libraries

As well as the wonderful traditional libraries found in most communities, **street libraries are popping up** in neighbourhoods all around the world, where you can swap a book you've just finished with something available there. It makes my heart sing when I'm out walking somewhere and I pass a street library.

We posted a picture on our 1 Million Women Facebook page not so long ago, showing an old double-door fridge, painted brightly and jam-packed full of books, sitting there among the trees on the side of a road. Now that's what I call a 'street library'.

Buy second-hand, then pass it on

When there's something you'll need for a few years – maybe a car, a bed frame for your child, a dining table – see what's available via an online marketplace first. Then, a few years down the track, **when you don't need the item anymore, offer it up to someone else.**

Overconsumption can find new levels of excess – outside of our normal patterns of behaviour – when big events punctuate our lives. It could be a wedding, a big birthday, a religious celebration or Valentine's Day. What happens to us around these 'big events', which are promoted by people who want us to buy more and more? Some of us unravel all our good intentions, and just go nuts on stuff.

Unfortunately, many of us turn into crazy consumerist people. But there are **heaps of ways to avoid these super-excessive times of year**, ideally without creating any waste at all.

I'm dreaming of a no-waste Christmas

During the festive season, we travel more, eat more, drink more and spend more than at any other time of the year. Australians spend an average of A$2500 each in the six-week lead-up to Christmas! Among all the good times, there's a huge impact on the environment.

In 2017, Australians spent A$2.4 billion on the Boxing Day sales, breaking records from previous years. The day is filled with crowds of people pushing into shopping centres to score crazy deals on things that they likely don't need – a perfect demonstration of overconsumption.

Christmas trees

Think of all those poor old post-festive trees being left out on the grass verges around the country: the Christmas-tree industry produces huge amounts of waste every year. **Try buying or renting a live tree, in a pot, which will produce oxygen and soak up carbon dioxide (plants do that).** After Christmas, ask your local nursery for advice on replanting your tree for next year. Don't fancy pine? Go fragrant with a rosemary bush, which will live for years and will be a useful addition to your kitchen pantry, or get creative with an upcycled or homemade tree.

The gifts

Put your hands to work by making homemade relishes and jams, homemade salsas or superfood balls, preserved cherries or mangoes, DIY raw hazelnut spread, home-baked cookies, your own beeswax wraps (see pages 236-7 for instructions), or sew some mesh bags … just to name a few of the things you can do!

Kids love making gifts, gift-giving and gift-getting, so make sure they hear you talking about why you're choosing to do away with gift-buying. The more we can instil in them the importance of a smaller ecological footprint, and the value of love and sharing over getting mired in consumerism, the better.

❗ Each year, we throw away enough wrapping paper to go **around the world nine times**.

The rubbish

Typically, we throw out 30 per cent more rubbish during the festive season, with the big culprits being decorations and wrapping paper.

Think about what's already around the house: consider wrapping presents in newspaper, fabric, furoshiki wraps (see page 275 for instructions) or even a decorative tea towel. If you're handy with a sewing machine, there are plenty of DIY ideas online using fabric scraps. As for decorations, 'buy once, buy well' is a good motto, so you can reuse them every year, or just make your own.

If you do receive something that's been gift-wrapped, open it carefully so you can use the wrapping again for someone else's gift. Or decorate cards with it. Or use it to cover books.

What about Easter?

In the UK, more than 8000 tonnes of waste is generated from Easter cards and Easter egg packaging every year. Seriously?

Choose eggs wrapped in foil only, then recycle the foil by scrunching the empty wrappers together (into a ball about the size of an apple) and placing them in the correct bin.

 75% of all aluminium ever produced is **still in use today** because of recycling!

And those pretty party balloons?

Sorry to burst your balloon, but these seemingly innocent party decorations are an environmental disaster.

To a sea turtle, these brightly coloured pieces of plastic have a striking resemblance to jellyfish or sponges. They are regularly ingested, often with fatal results. **It can take just a single fragment of a balloon to end the life of a sea turtle** and, with all species endangered or critically endangered, we need to do everything we can to conserve turtle populations.

Although some party shops will claim that their latex balloons are 'biodegradable', they're not a great alternative. Latex claiming to be 'biodegradable' still doesn't degrade for at least a few years and, in that time, can cause just as much harm to the environment as regular balloons. Let's throw flower petals instead.

The experience

The best thing about big events is the time you spend with your loved ones. So, why not make this a gift? Spend time, not money – or at least spend your money on an experience that the family can enjoy. Like simply being together!

While you're planning, make sure you let others know what your personal expectations are for the gifts you are about to receive.

Don't be afraid to tell those around you that you would prefer a donation to a charity or an experience rather than stuff.

DIY: Furoshiki – Japanese fabric wrapping

You'll need:

- A gift, homemade, hopefully
- A square-ish piece of cloth, ideally something you are reusing or found in an op shop:
    - Scarf, bandana or handkerchief
    - Tea towel
    - Offcuts from a sewing project
    - Any lovely fabric you can find!

Directions:

1. Lay out the cloth, orientating it like a diamond. Place the gift in the middle of the cloth.

2. Fold the top corner over the gift and tuck it underneath. Then do the same with the bottom corner.

3. Gather the left and right corners and tie a double knot, going left over right, then right over left. Secure the knot tightly.

4. Tuck in any overhanging corners. And you're done.

? Take a minute to think about all the purchases you make in one day, week or month.

Now consider: **are there any purchasing decisions you can change right now?**

14

Money toolkit

Women's economic power

For years, I had no idea that my money was supporting dirty coal but, when I found out, I was utterly shocked. It wasn't even on my radar until after I'd started 1 Million Women.

There I was, working hard to change my life in every aspect while, at the same time, I was funding the coal industry.

Women can wield a lot of power with our money. Same goes for men. But we need to ensure that our money is aligned with our values. Our financial decisions and choices can help to shape the kind of world we want to live in.

When you boil it all down, there are four key areas to money: spending, saving, investing and giving.

It's totally counterproductive if we care deeply for humanity, love our Earth, try hard to cut waste and pollution in our daily lives, then let banks, superfunds and other financial institutions use our money to invest in fossil fuels and other environmentally destructive industries.

The same goes for weapons manufacture, gambling, tobacco and other socially negative and ethically unsupportable sectors, which threaten the well-being of today's and future generations.

The good news is that we have choices

There is great news on the money front, so long as we make sure we're informed. **The financial aspect of our lives now has a lot more 'ethical investment' choices available than existed even a decade ago**, when 1 Million Women was getting started.

To seek it out, start with an internet search – but you need to know what to look for and ask the right questions.

Here are some key things you should consider:

- **The big one for climate change**: Will your money be used to support fossil fuels such as coal, oil and gas in any way?
- **Followed by**: Beyond fossil fuels, is any other environmentally damaging process involved?
- **Don't stop there**: Ask plenty of questions. Is it ethical and socially equitable, including gender equity and respect for diversity, workplace health and safety, fair trade and more?

Here's what to do:

- Choose a bank or institution that, categorically, doesn't loan to or invest in fossil fuels or other greenhouse-gas polluters and, better still, does invest in renewable energy and other socially responsible causes such as helping the advancement of electric cars, better recycling centres, public transport and health.

- Don't fall into the trap of thinking that investing ethically means you'll have to accept poor financial returns (the reverse can be true if you get good advice and make the right choices).

- Set high standards. How much better will you feel, knowing that your money is helping to support the growth of renewables, energy efficiency and other innovative activities that are making our world a better place?

✔ At 1 Million Women, we recently found out that one of the companies our superfund invests in **collects discarded fishing nets**, which would otherwise damage our oceans and trap marine life, and **turns them into carpet!**

So, not only do they *not* invest in coal, they *do* invest in **initiatives that align with our values.** How great is that!

What is my bank doing with my money?

When we deposit our savings in a bank, the money doesn't just sit there. It gets loaned and invested so the bank can make profits and pay you interest. Unfortunately, all too often, our money is being used to support industries and corporations that we wouldn't support ourselves.

If you don't want your money being invested in fossil fuels and other harmful industries, ever, then you need to act. **Your personal banking choice, with a little investigation, can make an enormous statement.**

It's a key action that most of us can take. One that will have ongoing positive effects for the climate, and that will send a powerful and definitive message to the banking world – their customers don't approve of investment in the fossil-fuel industry.

> ❗ 'Since 2008, the four big banks in Australia have loaned **$70 billion** to the **dirty fossil-fuel industry**.'
>
> – Market Forces website

Compare banks

There are banks out there that really do walk the talk about 'caring about their customers and the environment'. Do your research. Globally, there are excellent sources of reliable and up-to-date information on which financial institutions do ethical well, and which don't.

If you are in Australia, check out Market Forces, our go-to source for comparing your current bank with better options when it comes to ruling out fossil-fuel investment. They've got a long list of banks that invest $0 in fossil fuels; you can compare banks to determine which one's right for you.

I've decided to switch banks ... what now?

Put your current bank on notice that you're leaving, and make sure they know why. Before you close your old account, make sure to redirect your direct debits to the new bank first (get them to help you), and try to avoid making any new payments that will sit in the shady area that isn't your 'available balance' for a few days.

Your old bank will definitely be on the lookout for this kind of activity – they want to keep you any way they can. So, **expect them to try to talk you out of switching**. It's 100 per cent your decision whether you stay because they make it worth your while to do so, or whether you stick to your principles and move on, regardless.

Changing savings accounts will always be easier than shifting bigger banking relationships, such as home mortgages and business or personal loans. It's important to consider switching it all, though, if you've found a new bank that aligns with your values and doesn't invest in the fossil-fuel industry.

Remember, the financial industry is highly competitive, and there are many avenues to find expert advice and help.

Shout it from the rooftop

When you do switch to a bank aligned with your principles, you'll feel like a superhero. You'll be an absolute champion for making such a positive change.

Tell the world that you did it, why you did it and how easy the process was for you, so other people are inspired to follow.

I promise you, knowing you are with a bank that doesn't invest in dirty fossil fuels will truly make you proud.

Your superannuation (retirement or pension fund)

Superfunds invest your retirement savings in businesses, with the long-term aim of making your money grow. **But too many funds choose to invest in unsustainable industries.**

With so much of the world's retirement funds contributing to global warming, we need to know where our money is going, and what it's supporting. None of us want our money to be supporting the fossil-fuel industry, directly or indirectly.

1 Million Women surveyed 1000 women in our community to gather valuable insights into their attitudes around super and how we can use our money to create a better world.

Key results included:

- 90 per cent of participants supported their super being invested in renewable energy
- 80 per cent supported their super being invested in businesses that keep plastic out of the ocean
- But the survey also found that some people knew their fund wasn't ethical, but either found looking into ethical superfunds too confusing, or were worried about the fees being higher.

> **!** When asked, 9 out of 10 people expect their super to be invested ethically, yet a staggering **55% of superfunds invest in fossil fuels**.

> **!** Only **2% of superfunds** invest in climate solutions such as **renewable energy**.

These comments from our survey respondents help to illustrate that what we want from our banks or superfund, and what we get, can be very different things:

- 'I'm ashamed to say that with so much to think about, I've just gone with my workplace super, despite not knowing how they invest and despite having thought about the fact that I would prefer ethical investments.'

- 'Until recently, I had no idea what companies my super was invested in. Then one day, I logged into my online account to see how much super I had ... and there were all these names of big polluting companies such as global mining conglomerates, along with the amount of my money that was invested in them. I was horrified!

  I checked my options and changed to a "sustainable" option – no alcohol, gambling or mining companies. Then I checked my bank – they don't invest with fossil-fuel companies, either. **I don't want my money enabling companies like that to ruin our environment and society**.'

- 'Superannuation is not something I have thought consciously about, regarding ethical investment; however, **it is extremely empowering to have the realisation that the money I earn can support conscious businesses with minimal lifestyle adjustments**.'

- 'It is such an important investment tool that so many people don't utilise to its full potential. **Our dollar is our vote!** And to have tens of thousands of "votes" empowering what I believe in and work for is very powerful.'

What does ethical investing look like?

There are three key dimensions to ethical investing; the best investment funds in this space are likely to use all of them.

- **Negative screening**: Eliminating investments in unethical and unsustainable industries.
- **Positive screening:** Deliberately seeking out investment opportunities in ethical and sustainable industries.
- **Proactive engagement**: Using the power of investments to pressure industries to be more ethical and sustainable.

Which banks or superfunds fit with my values?

Here's a simple checklist to make sure your money is working for the good of people and the planet as well as for your own financial well-being.

☐ **Action 1: Ask the question.** Is my bank/ superfund lending to or investing in industries that don't match my values?

☐ **Action 2: Find out the answer.** Call your bank or fund, and ask them directly – they need to know that people like you care. Ask them about their ethical, responsible or sustainable policies. Ask them if they invest in the fossil-fuel industry.

☐ **Action 3: Take decisive steps** based on what you find out, with the addition of independent and expert financial advice wherever you require it (such as checking whether you'll face exit fees or other penalties for leaving, as well as assessing the performance of your potential new bank or fund). Your menu of actions can include switching to a new bank or superfund yourself, telling your employer that you want to move your superfund (and why), or staying where you are if you like what you've found out.

Once you've done the research and found out the answers, if switching your bank or fund is the right decision then it may be easier than you think. Often your new one will do most of the work for you.

☐ **Action 4: Tell your old bank or fund why you've left them** and, ideally, tell your new one why you're joining them. The old one needs to know why people are rejecting them, and the new one should know that its attention to being more ethical is being recognised and rewarded by people like you.

☐ **Action 5: Share what you've done with the world**, your family, friends and networks. It will get others thinking about their money, too!

Making the switch

Most of the time, we don't think about superannuation and, even when we do, we can end up pushing it to the bottom of the 'to-do' list.

But we need to try hard on this one because switching our super is a one-off action that will have a massive, lasting impact.

Once you've done your research, the actual process of switching to a superfund can take less than five minutes.

! Currently, on average, **men retire with A$166,000 in super**, while **women have only A$96,000** – that's about **42% less for women**, despite our longer projected lifespan.

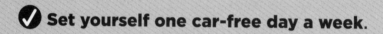 **Set yourself one car-free day a week.**

On that day, either don't travel anywhere, walk or use public transport.

15

Travel toolkit

Getting from A to B with the least impact on the planet

By good fortune (as opposed to smart planning), my kids' kindergarten and primary school was on our street, just a few minutes' walk away. Their high school is only in the next suburb, too. So, when I started looking into the environmental impact of our transport choices, we were already ahead on that front. But personally, I drove a big fuel-guzzling car, and I rarely caught a bus or train anywhere.

In my quest to live with the least impact on the planet, how I got from A to B was the last major issue I addressed. This was the hard one.

It needed a rethink of my life and some better planning (those of you who know me will agree – planning isn't my strong point).

> ✔ France and the UK will **ban the sale of new diesel and petrol cars** by 2040, and Germany is leading the way with a ban proposed for 2030.

But now, I catch public transport 60 per cent of the time. We are down to being a single-vehicle household. We walk around our local area a lot, use public transport constantly, jump in Ubers or taxis when we need to and, occasionally, rent or car-share a vehicle if a bigger trip is being planned. We're saving money and we're lightening the load on the planet.

This was a big lifestyle shift. I had to work hard at it but, once it became embedded in my way of thinking, it changed my whole outlook on how I get around. I do still drive when I need to, but **now I consider public transport to be my first option instead of my last**.

I know, not everyone has the same choices. If you live in the countryside, far from public transport, you need a vehicle. You'd be restricted and unhappy without it.

> ❗ Collectively, **cars and trucks** account for nearly **one-fifth of all US emissions**.

Every journey counts

Without counting the carbon emissions from initially manufacturing and transporting the car, running a mid-sized car creates about 4 tonnes of CO_2 pollution a year. **Around 8 per cent of Australia's total carbon dioxide emissions are from cars and light commercial vehicles,** so the way we get around makes a big difference.

To get yourself thinking about the journeys you make, ask yourself some questions:

- **If you have multiple cars, can you live with one?** If you have one car, can you drive it 50 per cent less? Or can you live without it? I know it might not always be an option, but it's worth asking the question.
- **Can you send your kids to school closer to home** so they can walk? If you have children or grandchildren at a nearby primary school or kindergarten, maybe you can set up a walking school bus. You'll keep fit along with cutting CO_2 pollution.
- **Can you work closer to home** or negotiate flexible arrangements? Or do more teleconferences, or even telecommute?
- **Can you switch to public transport** for most or all of your regular trips?
- **Is carpooling an option?** Most cars seat up to five people – by not filling it up, we are using an enormous amount of energy for a small amount of transport. Driving our cars solo is one of the most polluting things we can do.
- **Can you consider car sharing?** Look at sharing arrangements with friends, family and neighbours. If you drive less than 10,000 km a year, is car ownership really the best option?
- **Bike riding and walking** are not only simple, great ways to reduce pollution, they're also extremely beneficial to your well-being!

Drive down your CO_2 emissions

Sometimes, there's no way of getting around it – there's too many of you, with too much stuff, and the car is the best or only option. When you do drive, however, you can still do your part to help reduce CO_2.

Keep these in mind when you jump in the driver's seat:

1. **Avoid quick starts and aggressive driving.** Flooring the accelerator can produce high pollution rates and wastes fuel.

2. **Stick within or under the speed limit.** Studies have found that driving 90 km/h instead of 110 km/h on the motorway can **improve your fuel economy by about 15 per cent.**

3. **Avoid hard braking.** Especially when you're in traffic, don't slam on the brakes.

4. **Try not to idle for more than 60 seconds.** Idling your engine generates more pollution than driving, and it wastes fuel. If you're stuck in bumper-to-bumper traffic and you aren't going anywhere, switch it off.

5. **Open your windows before turning on the air conditioner.** On a hot day, removing the hot air through open windows before turning on the air conditioning will reduce your fuel consumption and nitrogen oxide emissions in some vehicles.

6. **Don't be too cool.** Air conditioning should be limited as it stresses the engine and uses more fuel.

7. **Avoid unnecessary trips.** Do you really need to go to the shops right now? Make a list, and go on your way home from work tomorrow. **Doing five errands in one trip makes a lot more sense than five separate trips.**

8. **Size matters on the road.** When buying a new car, choose one as small as you can for your day-to-day needs. And make fuel efficiency the top priority.

9. **Don't use a car for short journeys.** If your journey is less than a kilometre or two, then walk or cycle. It's good for you, your budget and the environment.

10. **Watch your weight, and lighten the load.** Reduce weight, and your car's fuel consumption, simply by removing unnecessary items from your car, which don't need to be there for a particular journey.

11. **Remove roof racks when not in use.** The wind resistance from roof racks dramatically increases fuel consumption, so take off the racks and luggage capsules when you're not using them.

12. **Tyre pressure makes a difference.** Ensure you have the correct tyre pressure. Every 6 psi the tyre is under-inflated increases the fuel consumption by 1 per cent. It's one of those little things that makes quite a big difference and only takes minutes at the garage to fix.

13. **Don't overdo it when refuelling.** Avoid overfilling the tank, as spilled fuel evaporates and releases harmful emissions.

> ✅ Every time you avoid
> **driving 5 km**, you save about
> **1 kg of carbon pollution**.

Holiday travel

Are you travelling somewhere this year? As you journey across the planet, think about how you can love Earth, too.

Travel is incredibly joyous, but it can also be highly polluting. **Plane travel alone uses large amounts of fossil fuels**, which creates greenhouse gases, and this is before you've even arrived at your destination.

Travelling responsibly and sustainably means planning ahead, giving back where you can, conserving natural resources, supporting local cultures and making a positive impact on the places we visit.

Here are some ways to help you travel with less impact on our planet:

1. Unplug your home or office. The first step of sustainable travel starts at home. The key word here is 'unplug'. Any appliances you won't be using – computers, printers, televisions, microwaves and so on – make sure they're all unplugged at the wall. Appliances can be massive energy hogs around the home, so remember to do this before you head out.

2. **Pack your environmental ethic.** When you're on holidays, you want to kick back, relax and take it easy, which, of course, is a wonderful thing to do. However, this doesn't mean you can forget your environmental manners! Keep to the practices you do at home, such as **turning off lights when you leave the room and conserving water.** If they don't employ this policy already, how about asking hotel staff to not change your sheets and towels each day because, really, how many of us do this at home?

3. **Ask questions before you book.** If you're booking a hotel, look for an environmentally responsible one. Always ask a bunch of questions, such as:

- What is your environmental policy?
- What sorts of policies have you implemented to reduce water consumption, conserve energy or recycle waste?
- Do you support any projects that benefit the local community?

4. **Stay simple.** Another option when looking for a place to stay is to choose the smallest, simplest option. **Smaller properties, with fewer amenities, consume less energy.** If you've ever stayed in a small, family-owned bed and breakfast or hostel, you'll know the experience is typically more personal and authentically local. It could be exactly the trip you're looking for and be more sustainable at the same time.

5. **Visit during a festival.** Sustainable travel is all about supporting local communities and cultures, and making a positive impact on the places we go. By scheduling trips in conjunction with **vibrant local celebrations**, you'll gain a richer experience of the culture, and it also helps to support traditional crafts and customs.

6. **Support sustainable tourism practices.** Similarly to hotels, choose tour operators that support sustainable tourism through their bookings and operating policies, and ask a lot of questions to make sure you are choosing the best one.

7. **Go paperless.** Use online resources to plan and book your trip. If you have a smartphone or tablet, **apps can be a wonderful travel resource to help you go paper-free**. There are so many maps and travel guides, including green-map apps that help you find the most sustainable experience of your chosen cities. Only when absolutely necessary, print out maps and other materials, at home and on recycled paper.

8. **Bring your own water bottle or coffee cup.** Waste is a pollution-intensive aspect of travel. A great way to cut down on day-to-day waste is by bringing your own water bottle and refilling it, rather than purchasing plastic bottles that then get thrown away. It's a good idea to get a dishwasher-safe bottle with a water filter, depending on the places you're going. If you're a coffee fanatic, then travelling with a lightweight, reusable coffee mug will help to reduce single-use waste. And don't leave home without your metal or wooden cutlery and cloth napkin set. Just remember to put it in your checked luggage if you're flying!

9. **Use alternative transports.** Try arriving at your destination by bus, train or ferry. Alternatively, if you do drive or fly, try to keep car travel to a minimum when you arrive. Walking and riding bikes can be fabulous ways to discover a place, as is public transport. These modes can provide a more authentic experience because you'll truly be among the place's daily life. **If you do use air travel, offset the emissions.**

10. **Lend a hand.** Make a positive impact on any community you visit by giving back in some way.

11. **Keep it local – explore wonders that are closer to home.** Getting there will cause less pollution than travelling further afield. Along with all our other tips, this would make for an extremely sustainable journey.

> ✓ As of 1 January 2017, every **electric train** in the Netherlands runs on **clean renewable energy**.

Jumbo polluters

Air travel is one of the toughest challenges for cutting carbon emissions because there's currently no replacement for high-polluting fossil fuel to power aircraft engines. While planes are becoming more fuel efficient, more people are also flying, which means that the only way to deliver real reductions in air-travel emissions is to fly less often.

Offset when you fly

When you do need to fly somewhere, use airlines that offer offsetting programs.

Travel with purpose

Don't just fly somewhere to tick it off your list, or buy into packaged holidays to see the 'unmissable sights'.

Think not only about *where* you want to go, but *why* you want to go there.

Try to place the focus not just on what you need from the holiday but also on what you can give back to the destinations and local people. This is a more holistic approach to travel, and is sustainable both for your mind and the environment.

Driving into the future

Electric cars are a way of securing a sustainable future for our global transport system. Even if a large portion of electric cars are currently powered by coal power plants, it won't stay that way for long. Everywhere across the world, renewables are becoming cheaper, more viable sources of energy.

So, as more and more cities and towns make the transition from fossil fuels to renewable energy, **it will be vital to have more electric cars on the road**. This is a very exciting story and one that we'll be part of.

! With transport responsible for around **18% of Australia's carbon emissions**, converting to electric vehicles could reduce emissions by at least **15 million tonnes** by 2030.

A million thanks

My beautiful friend Michelle Grosvenor has been there from the beginning, before we even conceived of 1 Million Women. Her experience and wisdom guided me, the newcomer, as I took my first tentative steps.

Michelle's incredible support soon multiplied as we put together an advisory board of wonderful women, who all gave freely and generously of their time and advice to get 1 Million Women off the ground: Wendy McCarthy, Sam Mostyn, Anita Jacoby, Kim McKay, Bernie Hobbs, Rosemary Lyster and Ilse O'Reilly. And the blokes, too: Michelle's partner, Paul Gilding, and mine, Murray Hogarth.

It was real pioneering stuff, those early days of 1 Million Women: The Women in Climate Change breakfast series, Our Recipe for Change big weekend, The Save Summit. I couldn't have done any of it without you all. You are forever a part of the story of 1 Million Women.

More women stepped up to serve on our Board of Directors, providing the support that not-for-profits need and giving selflessly to help 1 Million Women flourish: Peri Hunter, Caroline Pidcock, Nicki Hutley, Margot McNeill, Megan Coombs, Kirsty Ruddock and Carolyn Pridham.

I have so much gratitude for all of you, and for the dozens of women who stood with us on our advisory team as well as 1 Million Women Ambassadors.

1 Million Women would never have existed, nor continued to flourish, without the generosity of our donors and funding partners. There are so many of you, I can't name everyone. I thank you all for believing in our mission.

Thank you to Anouk Darling and the team at Moon, the talented agency that brought 1 Million Women to life as a brand. Those late nights brainstorming, that red wine. Your creative brilliance, generosity of spirit and your time, all done pro bono.

Thank you to Andreas Smetana for your magnificent artistic vision, turning an old Aussie hit by the king of pop into our anthem for climate action and hope, together with The Passion Collective, the girls from Aim, the big voices of Wendy Matthews, Deni Hines, Melinda Schneider and Ursula Yovich, and a whole bunch of fabulous women.

I am so lucky to work with a passionate team who share our vision and commitment to changing the world. Some have moved on to other roles, and all have made 1 Million Women braver, bolder and better – Grace Liley, Tessa Marano, Eva Davis-Boermans, Harriet Sparks, Bindi Donnelly, Holly Royce, Bronte Hogarth, Shea Hogarth, Laura Oxley and Barbara Dick. I also thank the constant stream of interns and volunteers who join our team and give so much. 1 Million Women thrives on their fresh, enthusiastic and dedicated inputs.

Thank you to Naomi Chrisoulakis and to Fiona Daniels for helping me to tell my story.

And to our community of women and girls: you are 1 Million Women! You are inspiring. Thank you with all my heart for making 1 Million Women the movement of many voices that it is today.

In loving memory of my 1 Million Women partner Tara Hunt.

Natalie

Be part of the 1 Million Women movement

- Download our free 1 Million Women app – search for 1 Million Women in the Apple and Google Play stores (it's scheduled to launch near the end of 2018)
- Join up on our website and be counted – 1millionwomen.com.au And take part in any of our campaigns.
- Join our social media community – Facebook @1millionwomen; Instagram 1millionwomen; and Twitter @1millionwomen
- Become a 1 Million Women Ambassador. We need you. For more info, contact us at ambassador@1millionwomen.com.au You can access online slides and videos, and a guide for how to hold your own 1MW workshop with your friends, family or community, to give others the inspiration to profoundly change the way they live. It's all free.
- Send us an email at enquiries@1millionwomen.com.au